# NATURAL GAS BY SEA

First Published 1979
Second Edition 1993


© Roger Ffooks


ISBN 1 85609 052 3


Illustrations and design by
Brian Roll

Printed and Published by

Witherby & Co. Ltd.
32-36 Aylesbury Street
London EC1R 0ET

Tel No: 071 251 5341
Fax No: 071 251 1296

British Library Cataloguing in Publication Data
   Ffooks, R.
   Natural Gas by Sea 2nd Edition
   The Development of a New Technology
   ISBN 1 85609 052 3

# NATURAL GAS BY SEA

## The Development of a New Technology

### Roger Ffooks

WITHERBY · LONDON

*This volume is dedicated
to all those whose skills,
sense of purpose
and perseverance
have enabled the Technology
of Liquefied Natural Gas
by Sea to become a reality.*

# About the Book

Forty years ago the first attempts were made to find the materials, and to establish practical design and building techniques, for shipping liquefied natural gas (LNG) at its boiling point of minus 165° Centigrade safely across the oceans of the world in commercial quantities.

Many experts pronounced it impossible — or dangerous — or both, but the perseverance of a small team of engineers in the U.S.A., encouraged by the British Gas Industry, proved that it was at once possible, safe and commercially viable.

Europe accepted the challenge and developed their own technology which was to compete with, and quickly over-take, the U.S.A. in the race to participate in this new and potentially lucrative trade.

The book describes the way in which this fascinating new technology was developed and how, within a relatively short time span, it became a well established, highly sophisticated and safe, branch of modern marine engineering; ships of up to 135,000m³ capacity and, for their size, the most expensive merchant ships afloat, are now in regular service worldwide. Indeed, today it is quite difficult to conceive of life without LNG. No small part of the fascination of this story is the atmosphere of commercial secrecy in which the technology grew and prospered, and the wide variety of designs which has evolved.

The technology is still developing and the trade expanding. This new edition brings the reader up to date and offers an 'educated guess' as to what the future might have in store.

# The Author

Roger Ffooks, B.Sc., F.R.I.N.A. has been involved with gas tankers, both LNG and LPG, since the early 1950s, first as senior Naval Architect with Shell International Marine Limited and subsequently as Technical Director of Conch Methane Services Limited.

He was directly concerned with the design, construction and operation of *Methane Princess* and *Methane Progress,* the first two commercial LNG carriers, and subsequently the development of one of the two basic membrane designs. He contributed to the preparation of the IMO Gas Carrier Code as representative of the International Chamber of Shipping and the British Delegation and has written a number of papers on LNG ship design and development.

Roger Ffooks left Shell in 1976 to set up as an Independent Consultant, primarily in the gas ship field, based in London; he is now "semi-retired" and lives in West Dorset.

# Contents

# Illustrations

Key to Cover Illustration

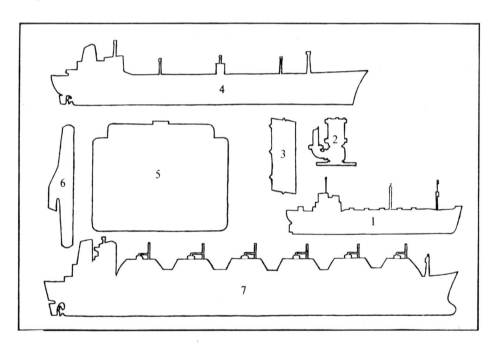

1. *Methane Pioneer,* the first prototype LNG carrier
2. Electrical submerged pump
3. Cross section of the insulation membrane of the Technigaz Mark I system
4. *Methane Princess,* the first commercial LNG carrier
5. Cross section of the Gaz Transport membrane ship
6. Equatorial ring extrusion for the Moss sphere supports
7. Profile of a 125,000m³ LNG carrier with spherical tanks

**Colour Plates**
opposite page 45
*MV Methane Pioneer*

opposite page 56
*Beauvais*

following page 134
Lake Charles fire tests, and Maplin Sands Spill Tests
*Jules Verne* at Le Havre — 1965
*Methane Pioneer* at Lake Charles — 1959
*Mostefa Ben Boulaid* — 1976
*Methane Progress* on trials — 1964
*Arctic Tokyo* loading in Alaska — 1969
The Conch 125,000³ design, as offered by Sumitomo shipyard
25,000m³ sphere in transit to General Dynamics shipyard
*LNG Aquarius* on cold trials at Canvey Island Methane Terminal

# Foreword

Seaborne transportation of liquefied natural gas (LNG) is a young industry. Young enough for many of those who are still active in this fascinating segment of international shipping to remember very well the first searching steps to find a solution to what was a major technical challenge.

A number of developments have transpired since the first edition of "Natural Gas by Sea" was published in 1979, and with the continually increasing interest in natural gas as the environmentally preferable fossil fuel, it is timely that Roger Ffooks has taken the opportunity to bring us up to date on what is happening in LNG shipping.

The revision provides some fascinating insight into how the different designs for this most advanced marine technology were developed and the intense rivalry between the developers and licensors of the competing technologies, frequently with the gas conferences as their battleground. It was a time of achievement and reward, but also frustration and disappointment when progress was slow.

The potential danger in handling a liquid at these ultra low temperatures and the huge investment necessary to implement an LNG project demand a total commitment to safety and reliability. This book demonstrates how this has always dominated the thinking of the people involved and how special safety features have been incorporated in the design of these ships. The importance of safety and reliability and the recognition that the failure of one operator would be a serious setback to all the others was the background against which the Society of International Gas Tanker and Terminal Operators (SIGTTO) came into being in 1979.

It is vital that the outstanding safety record of LNG shipping is maintained in order to permit the considerable increase in the size of the LNG fleet which will be required to serve the world's increasing need for this clean source of energy.

The book provides a valuable introduction to LNG transportation technology for students as well as professionals and is highly recommended reading for anybody who would like to understand the problems of seaborne transportation of liquefied natural gas.

Alf Clausen
President
SIGTTO

# Preface
# and Acknowledgements
# to first edition

Ever since LNG ship technology finally became 'established', it has seemed that someone should record the story of its birth and growing pains. To one who has been involved in one way or another for almost the whole period of its development and therefore lived with its excitements, hopes and disappointments, the fascination and satisfaction of helping to break new ground, it needed only a little encouragement to make the attempt.

Being very conscious of my limitations I can only hope that it has been possible to convey in this book an impression of the atmosphere in which the technology grew and an adequate idea of some of the problems involved and the people who shared in their solution; of the people, I can only say that it has been a great privilege to have worked alongside them or to have met them as commercial adversaries – either way I owe a debt to the development of LNG ship technology for the development of very many friendships.

Students of the technology will find references to fill the many gaps in these pages. I have made no attempt to write the definitive text book; it would have been dull – nearly all text books are dull – and LNG has never suffered from that characteristic.

In the process of collecting material or refreshing my memory I owe thanks to a great many people; in particular, I would like to thank Nigel Bruce, Ian McCallum, Tony Rogan, Chuck Filstead, Mike Robinson, John Sommerville and André Chemereau, all of whom allowed me to rummage through and extract information from their files and papers. Special thanks are due to Sir Denis Rooke, Chairman, British Gas; Audy Gilles, Managing Director, Gaz Transport; Bob Young, Chairman, American Bureau of Shipping; Jim Henry, Charlie Zeien, Monte Banister and Al Schwendtner of the J. J. Henry Co. Inc.; Jean and Jacques Kohn, Presidents of Baltek; and Dan Kilmer, President, J. C. Carter Company; each of whom spontaneously made available that additional, tangible, support so helpful at the start of a project.

To Mr William Wood Prince I am grateful for permission to quote at length from one of Willard Morrison's early reports; to Brian Singleton many thanks for reading proofs and his many useful suggestions; to Lorna Gentry, Christopher Burness and Brian Roll my appreciation of their patience and co-operation. Finally, my thanks to Barbara Rennie, my secretary and assistant, without whose encouragement I would never have been brave enough to start and without whose hard work, and moral support in moments of despair, I would most certainly never have finished.

# Preface
# and Acknowledgements
# to second edition

Some of those who helped in the first edition of this book have retired from 'active service' in LNG — some, including Jim Henry who contributed so much to the technology in its early days, have sadly died.

Others, happily, are still active and have continued to help with the updating of this 'history'; in particular Everett Hylton of Cryodynamics Co., Dick Chadburn of SIGTTO, Bill Wayne of Shell International Marine, Dr. Fujitani, IHI and Brian Singleton who once again read through the proofs with his usual efficiency. And of course my thanks to Barbara, now my long-suffering wife, for translating my dreadful handwriting into a legible form.

# Abbreviations used in references which appear at the end of each chapter

ATMA            Association Technique Maritime et Aéronautique, Paris.
SNAME           Society of Naval Architects and Marine Engineers,
                New York
RINA            Royal Institution of Naval Architects, London
LNG 2 (etc)     International Conferences on Liquefied Natural Gas
                (sponsored by International Gas Union, International
                Institute of Refrigeration and Institute of Gas
                Technology)
Gastech 74 (etc) International LNG/LPG Conferences, organized by
                Gastech Ltd., England
                (predecessors: LNG/LPG Conference, London, 1972
                           : LNG 73 – Second International LNG
                             Conference, London, 1973) – now
                             Gastech Rai Ltd., England

# Periods covered by Chapters and Some Landmarks

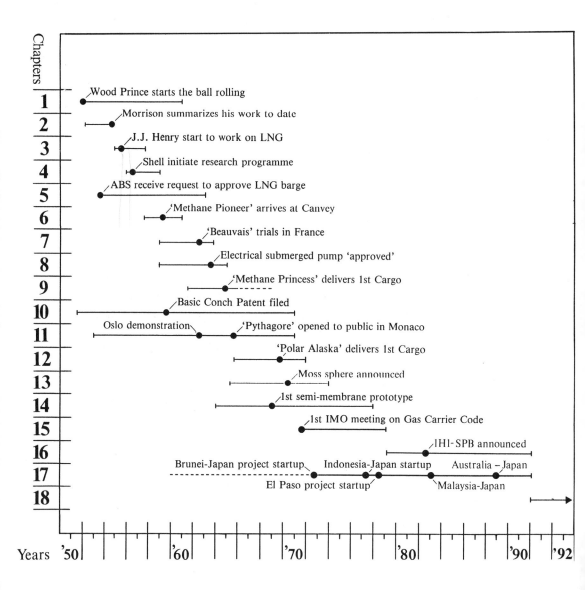

| Chapters | |
|---|---|
| **1** | Wood Prince starts the ball rolling |
| **2** | Morrison summarizes his work to date |
| **3** | J.J. Henry start to work on LNG |
| **4** | Shell initiate research programme |
| **5** | ABS receive request to approve LNG barge |
| **6** | 'Methane Pioneer' arrives at Canvey |
| **7** | 'Beauvais' trials in France |
| **8** | Electrical submerged pump 'approved' |
| **9** | 'Methane Princess' delivers 1st Cargo |
| **10** | Basic Conch Patent filed |
| **11** | Oslo demonstration · 'Pythagore' opened to public in Monaco |
| **12** | 'Polar Alaska' delivers 1st Cargo |
| **13** | Moss sphere announced |
| **14** | 1st semi-membrane prototype |
| **15** | 1st IMO meeting on Gas Carrier Code |
| **16** | IHI-SPB announced |
| **17** | Brunei-Japan project startup · Indonesia-Japan startup · Australia – Japan |
| | El Paso project startup · Malaysia-Japan |
| **18** | |

Years '50  '60  '70  '80  '90  '92

# 1
# How It All Started

It is doubtful if any major technological development or, to use current terminology, 'breakthrough', can be credited to any one man or any specific point in time. There is no specific day when 'natural gas by sea' was conceived, nor one single person who conceived it – no Archimedes in his bath, or Newton reclining, pensively, under the apple tree. And who can put a date to either of those two discoveries?

Developments in the twentieth century tend to be the result of a gradual build-up of pressures, from different directions and for different reasons – most often a combination of commercial and political factors; furthermore, this combination is both complex and subtle, and the point at which it eventually throws out a specific piece of technology is invariably undefinable.

The period which leads up to this hazy pressure point may be numbered in years, rather than days or months, and is inevitably sprinkled with the ideas and quotations of the 'thinkers' of the world; but sooner or later there will appear both a thinker and a 'doer', and so it was in the case of Natural Gas by Sea: in fact, as will be seen, one followed another in surprisingly quick succession and, since by definition the atmosphere was favourable, the seed germinated and grew at a remarkable rate.

The combination of influences which provided the stimulus for the sea-borne transport of natural gas were several. First, the host countries' growing disenchantment with the oil companies' practice of flaring the 'unwanted' associated gas at their oil fields. This practice was widespread – and the reaction equally so; even in the late 1940s and early 1950s it was appreciated that energy, and therefore income, was being wasted. These so-called political pressures began to stimulate studies into ways and means of recovering or conserving this wasted gas; one thing was certainly very clear – there was no local outlet for it. Reinjection back into the reservoir was one solution, but practical or economic in only a few cases; piping it to the nearest industrial consumer was another – but most fields were too remote from suitable markets for this to be economically or politically viable. Transporting it by sea was the third and perhaps the least attractive solution at the time since no known technology was available to achieve this end.

The second main influence was commercial and ran on parallel lines to the first; certainly, if the technology *could* be developed by which the flared gas could be collected and sold at a competitive price there would be a commercial interest in so doing.

Both these influences were by now being felt strongly by the oil companies operating in the great oil producing areas such as the Middle East and Venezuela. In the USA however – the only really substantial oil/gas producer and consumer – a network of natural gas distribution lines was already well developed and liquid natural gas storage units were being built in order to provide for peak load demands. Here the commercial pressures took a somewhat different form – increasingly high prices to the consumer.

In Europe, and most particularly Great Britain who at that time had no indigenous source of gas nor any prospect of it, industrial expansion and demands for home heating were combining to stretch the nationalized gas industry's production capabilities to the limit.

It was in these, perhaps rather oversimplified, circumstances that the first 'doer' appeared in the form of the Chairman of the Union Stockyards of Chicago, William Wood Prince, who was convinced that it should be possible to liquefy gas (a known technology) which he could obtain cheaply from the Louisiana oil/gas fields, to store it in land based tanks (also a known technology), to transfer it to barges and transport it to Chicago where it could be stored and regasified; the missing link, river transportation, could surely be solved in a practical and economic manner? This was 1951.

By early 1952 Wood Prince had put his money where his convictions were by commissioning Willard S. Morrison, a consulting engineer of proven experience and ingenuity in the refrigeration field, to head a small group to carry out an urgent study of the feasibility of such a project. Morrison's terms of reference were quite unequivocal as he described them in an interview with

C. I. Kelly some years later:

> 'My assignment consisted of having to formulate a comprehensive scheme, provide the design of all the necessary equipment which could be manufactured with the least capital investment and operated at the lowest cost and with the *maximum of safety* at every stage.'

This was positive action – in fact the thinking had started as long ago as May 1915 when a Godfrey L. Cabot had patented an idea for '. . . handling and transporting liquid gas' – by river barge.

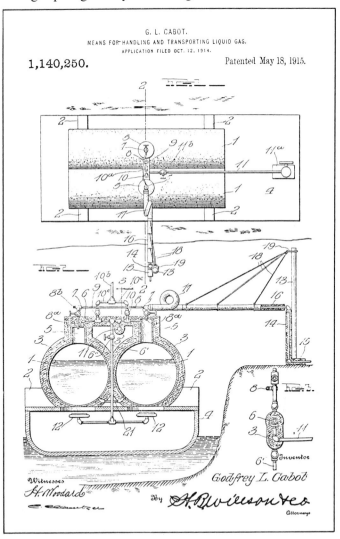

Fig. 1. CABOT's patent – 1915.

There is no record of any such craft having been built but the idea was there and indeed, as will be seen later, Cabot had in fact anticipated many of the basic characteristics of the present-day gas carriers.

In Europe, positive action was not taken until three or four years later when reports of Morrison's activities in the USA stimulated studies by shipowners and two major gas companies.

In Norway, by 1954 Dr. Oivind Lorentzen had developed and obtained Norske Veritas approval in principle for a spherical design of 17,000 ton cargo capacity; details of his patent and early ship layout appear in later chapters.

In the UK, also during 1954 and 1955, the Consulting Naval Architects, Burness, Kendall & Partners (now Burness, Corlett) had been commissioned by both Westinform and Wm. Cory to carry out a design study for a methane transport ship of about 14,000 tons – cylindrical (either horizontal or vertical) tanks being favoured, and designs conceived by the late Sir Denistoun Burney – a prominent British inventor of the time – being injected for good measure.

By 1955 the Shell Group had initiated a programme of work in London, The Hague and Amsterdam to consider ways and means by which LNG could be carried on board ship. Their Marine Department in London had the responsibility for the overall ship design concept, whilst the Amsterdam laboratories evaluated suitable materials – most particularly insulation. The project was co-ordinated in London, but no sooner had their efforts begun to make progress than they were abruptly stopped in 1956 as part of a severe cut back on all 'non-essential' R & D work following the outbreak of the Suez War.

Throughout this period the Gas Council of Great Britain, assisted by the senior engineering staff of the North Thames Gas Board, had maintained close contact with all current methane ship design developments, sponsoring detailed evaluation studies of European designs and visiting the Morrison group in the USA in August 1954 to make an on the spot evaluation of the progress there.

In France, the stimulus to investigate the marine transportation of gas took a somewhat different form – in that it stemmed, not so much from pressures to recover flared associated gas, as from the discoveries of large unassociated gas reserves in Algeria.

In 1954, Gaz de France commenced an in-house study of the feasibility of importing this gas into France – either by pipeline or by ship. By 1956 they had submitted a complete and detailed report to their Government who, after due deliberation, favoured the shipping solution. The commercial arguments were sufficiently forceful for the French Government to authorize detailed engineering studies – despite Suez, allocating to the Worms Group the responsibility for developing the ship technology.

The solution which was developed by Worms is discussed in more detail in a later chapter, but an incident which occurred in the interim must be mentioned here because it has a significant bearing on the direction in which LNG ship technology developed during the next ten years.

In June 1959 the Chairman of Worms et Cie, M. Labbé, visited the United States to attend the 5th World Petroleum Congress conference in New York; by that time it was clear to him that although considerable progress had been made in France, economies could well be achieved by taking a licence from the US group which by now had successfully shipped several cargoes of LNG across the Atlantic in the prototype ship *Methane Pioneer*. The reception he received was polite but uncompromising: '. . . it is impossible for anyone to compete with us . . . we are granting no licences'. Perhaps this inflexible attitude was due to the fact that the Americans were already negotiating a partnership deal with Shell, but whatever the reason the French reaction was predictable: '. . . then we will make our own study'. Methane Transport was formed, comprising a wide spectrum of French industry including Worms, Air Liquide, Gaz de France and Gazocean; no shipyard was included in this company, but that they were very closely involved in all future development work – and to their considerable benefit – will be seen.

Thus, during the 1950s, intensive fundamental research and development work was being carried out on both sides of the Atlantic – quite independently – with the USA having a good head and shoulders start. But the amount of reliable published information on LNG ship technology available during this period could be written on a single page of paper. The stakes were high and the players were keeping their cards very close to their chests indeed.

# 2
# First Attempts
# at a Solution

Willard Morrison's general terms of reference were to examine the feasibility of liquefying the gas available from a number of wells in the Mississippi Delta and to store this liquid prior to shipping it by barge up the Mississippi River to Chicago. In Chicago the gas would be stored prior to its regasification and used as fuel for the Stockyards cold storage plant – in addition the 'cold' would be available for the same purpose.

It was planned that the liquefaction plant and storage would be barge mounted – in order for it to be capable of moving from field to field, and logically a similar barge design could be used for both storage and river transport.

The design of the liquefaction plant presented relatively few problems except insofar that its capacity was well in excess of the peak load shaving plants then in existence in the USA: it was a question of scale – the material and design techniques were available.

The barges, however, presented a novel problem since not only was 'conventional' land storage not readily adaptable to water-borne craft but such tanks were themselves in a state of flux after the Cleveland incident of 1941 when a relatively newly built LNG storage tank constructed of $3\frac{1}{2}$ per cent nickel, low carbon, steel had failed catastrophically, causing considerable local damage and loss of life.

Morrison's team therefore decided to start from first principles, commencing with a comprehensive review of all possible materials suitable for safe operation at −165°C, the boiling point of methane. These could be conveniently divided into two general categories:

(a) materials of construction–for tanks, piping and equipment;
(b) insulation–load bearing and non load bearing.

In the first category came such well known materials as stainless steel, aluminium, aluminium bronze, copper and some of the precious metals such as silver. Of these only the first two were likely to be available in commercial quantities at remotely realistic prices–but there was little or no experience of the use of aluminium in larger structures and its welding was known to present problems.

As far as steels were concerned, it was well known that, by increasing the nickel content, the low temperature properties of the resulting alloy could be improved–indeed stainless steels were already widely used in the cryogenic industry; steels with lower percentages of nickel were also in current use at 'higher', but still sub-zero, temperatures. In the early/mid fifties, the newly developed 9 per cent Ni alloy steels had had limited use ashore, but certainly not enough experimental or service experience to justify their consideration for large scale shipboard application. At any rate the phenomenon of brittle fracture was not yet well enough understood, nor was there sufficient correlation of laboratory tests with service experience, to enable designers to feel secure with anything other than 18.8 stainless steel.

For insulating materials the choice appeared to be limited to mineral-based powders, fibrous materials such as glass fibre, with wood as a potential load-bearing material.

The overall concept of an insulated tank containment arrangement seemed to resolve itself into the choice of two clearly divided alternatives:

(a) a tank constructed of a material able to operate at LNG temperatures, outside which must be fitted sufficient insulation to protect the ship/barge structure and control the rate of evaporation of the cargo.
(b) a tank constructed of a material having no special low temperature properties, inside which must be fitted insulation capable of containing the liquid and protecting the steel from the low temperature of the cargo.

In either case it was judged advisable to design the tank to be as free from undesirable stress concentrations as possible – leading to the use of cylindrical or spherical shapes.

Both alternatives presented considerable design problems. The first (a) required that the tank be able to expand and contract freely and at the same

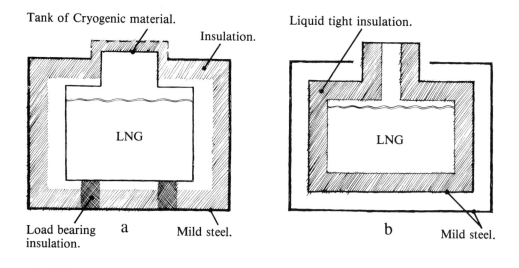

Fig. 2a. Alternative concepts for LNG containment.

time be securely located in the barge. The insulation had to be able to accommodate this movement if it were attached to the tank. The second (b) required very careful attention to the insulation system design to ensure that it re-remained liquid-tight for the whole of its working life – but it would mean a considerable cost saving in not using cryogenic structural materials, the welding techniques for many of which had yet to be established.

After due deliberation Morrison decided to pursue the second alternative based on the use of wood as an insulating material. Wood was, he felt, a readily available material which could be worked and assembled to reasonably close tolerances by existing techniques. The low temperature properties of available woods, however, were not known and the Forest Products Laboratory of the United States Department of Agriculture was therefore commissioned to carry out a comprehensive series of tests to establish the low temperature properties of three commonly used woods, Douglas fir, Sitka spruce and at FPL's suggestion, balsa. As FPL stated in the introduction to their report on this work (February 1952):

'Little work has been done on evaluating the strength of wood subjected to subnormal temperatures, particularly at temperatures lower than $-70°$F. Harry D. Tiemann pointed out in a series of articles in the *Southern Lumberman* beginning June 15, 1944, that "the question of temperature effect resolves itself into two phases: (1) when subsequently brought back to normal temperature, and (2) when used or tested at the temperature in question . . . however very little test work has been done under the second phase, mainly due to the difficulties in making the tests and in controlling the moisture condition, for a change in temperature is sure to cause a change in moisture equilibrium unless due precautions are taken." In addition, little work had been done on wood at subnormal temperature because the demand for strength properties has not been sufficient to warrant an extensive program of tests. The use of wood to serve as a heat insulator and as a structural support for tanks or vats containing a liquid at about $-260°$F now requires a better knowledge of certain strength properties in order to make possible the sound engineering design of such supports. It is necessary, therefore, to determine the temperature effect when the wood is used at the temperature in question.

'A co-operative agreement was arranged with a commercial co-operator, under which the Forest Products Laboratory agreed to build a test chamber capable of maintaining temperatures at about $-300°$F, and to make tests on wood at this temperature in flatwise compression, flexure, toughness, and linear thermal expansion. It was decided to conduct tests on three species of wood, lightweight balsa, Sitka spruce and Douglas-fir, since the co-operator desired that data be obtained for woods that are readily available but that, at the same time, would have high strength-to-weight ratios and low heat conductivity. A test chamber was constructed so that it could be cooled by liquid air, and so that only the test specimens, their supporting jigs and the loading heads were subjected to the subnormal temperature. Instruments for observing the loads and deformations were located outside the cold temperature zone. No attempt was made to control the humidity at this low temperature.'

These tests concluded that:

(1) the $-300°$F temperature had the effect of increasing the flexural and compressive strengths 40–400 per cent for various properties;
(2) the toughness values were increased up to 100 per cent;
(3) the coefficient of linear thermal expansion from $+80°$F to $-300°$F agreed with established values.

The density of the balsa samples tested were $4\frac{1}{2}$ lb per cubic foot.

The results were encouraging and thus it was on 12th March 1952 that the Balsa Ecuador Lumber Corporation in New York, the main US importers of balsa for model aircraft kits and door panel cores, were astonished to receive the following letter from Willard L. Morrison of Lake Forest, Illinois:

'Gentlemen:

I have in process of design, equipment, one phase of which will require balsa wood in considerable quantities if it is available for delivery in a reasonable time and if its cost is found to be economical for the application. Our design contemplates the use of balsa in thicknesses of 12 inches or more and lengths of 14 feet. There will also be uses in smaller size structures.

The balsa structure is being considered if suitable for constructing into a tank or tub, about 50 feet in diameter with 12 inch thick walls or staves and with a depth of tank of 14 feet. You will appreciate that lengths of balsa up to a foot or more thick would be desirable but that a design could be developed that could incorporate balsa lumber of other dimensions.

The nature of the process in which the balsa may be used is such that it is desirable to have a wood of the least density available; our tests for the properties of balsa so far having a density of approximately .08.

The quantity of balsa required if it is found to be an economical and available material would approximate that required for 50 or more tanks of the type described in the foregoing with balsa covering for the bottom and the top. I am also much interested in balsa wood having a thin veneer liquid impervious surface and shall appreciate any information you can send me regarding the availability and suggestions regarding such types of composite balsa board. Also I shall appreciate receiving all the information, publications and references that you can send me regarding the use and properties of balsa wood.

As a matter of possible interest to you there is enclosed a report on the properties of wood including balsa, at very low temperatures, tests which I had conducted in co-operation with the United States Forest Products Laboratory at Madison, Wisconsin.

Very truly yours,
Willard L. Morrison'

Subsequent discussions as to the prospective market indicated a vast potential order if the project was successful. Fortunately balsa trees grow fast and profusely in Ecuador so they were able to report that there would never be a problem in providing any foreseeable requirements for timber, this latter being, of course, an important consideration.

Within a year a prototype barge was under construction at the Ingalls Shipyard in Pascagoula, Mississippi with five vertical cylindrical tanks, to be internally insulated with balsa; it had a double hull to protect the tanks in the event of collision.

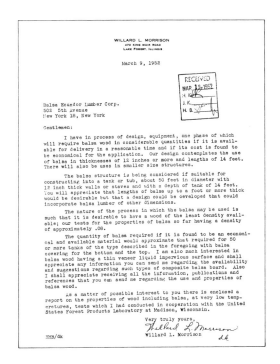

Fig. 2b. Willard Morrison's letter to Balsa Ecuador Lumber Group.

In their letter of March 1953 to the American Bureau of Shipping, in which the builders sought classification approval for this, the first LNG carrier, Ingalls explained that . . . the liquid methane cargo barge was intended for river service; it was to be used for the carriage of liquid methane or marsh gas at a gauge pressure range of zero to a maximum of six inches of water. In effect therefore, the cargo tanks were to be vented to the atmosphere. The density of the liquid methane to be carried inside the insulated tanks would be

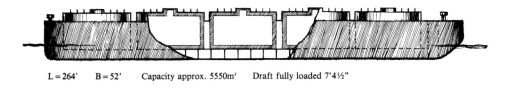

L = 264'    B = 52'    Capacity approx. 5550m³    Draft fully loaded 7'4½"

Fig. 2c. MORRISON'S LNG barge—*Methane*

17

only 26.5 lb/cubic foot. In full loaded condition the estimated draft would be 7 ft 4½ ins. Further . . . to make it easier to pass under the bridges, considerable ballast capacity was provided. The Bureau were also requested to agree to the barge being combined with another barge, a separate bow and a square-ended diesel towboat, to form an integrated tow about 600 feet in overall length.

By mid-1954 Morrison was able to summarize his work, much of the essential features of which were now covered by Patents, as follows:

'TRANSPORTATION

*Insulation*

The design and method of application of a structurally strong insulation for the inside of the metal walls of a storage or transport tank. This insulation having the strength, modulus of elasticity, coefficient of thermal expansion and other physical properties including the coefficient of heat conductivity, such as those of balsa and other woods, which will permit prestressing for the purpose of limiting stresses obtained in service, within the elastic properties of the insulation at temperatures down to those of methane, liquid air or nitrogen (258°F to 320° below zero F).

'This insulation also has the porosity and hygroscopic properties which permits prestressing by moisture content control. The porosity also provides an insulation, thus forcing the generated gas to discharge through the pores of the wood into the liquid within the tank and thereby preventing the outer metal wall from cooling appreciably.

'The application of heat transfer apparatus in an insulated tank containing liquefied natural gas, the elements of the heat exchangers of this apparatus being located both above and below the surface of the liquid in the compartment in which the heat exchanger itself is located, for the purpose of adding heat to boil the liquid and superheat the vapour, thus permitting the discharge or transfer of the liquid gas cargo in the tank in a gaseous phase for the purpose of emptying the tank, or for the purpose of increasing the rate of evaporation in the tank, when more gas is needed in a transport tow or vessel for propulsion, or for other power and heat purposes. The fluid heat transfer liquid flowing through the elements of this heat exchanger having the physical properties that enable it to remain liquid at the temperature of liquid methane (258° below zero F) at atmospheric pressure and also having the physical property of solidifying without increase in volume; the source of heat for the heat transfer fluid being derived from refrigerating brine, ordinary river or other water, or even from air.

'The design and application of the above described heat transfer system for the purpose of removing heat from the gaseous and liquid content of a liquid natural gas tank in order to control or prevent the discharge of gas from the tank without appreciable increase in pressure, the design and arrangement being such that the direction of the flow of gas and liquid in the tank through the heat transfer elements in the tank is reversed from the condition when heat is being added; the source of the refrigerating effect being from a separate refrigeration cycle mechanism.

'The design and application of a free running gas expander turbine driven fan, to a tank or transport vessel that has been insulated for liquid natural gas. This turbine being supplied with gas discharged from the safety or relief valve of the tanks and proportioned so as to mix sufficient air with the vented gas that the resultant mixture is too lean to ignite, also the provision of a "kick off" impulse for the aforementioned turbine that is automatically supplied from a tank, or other source of compressed air, that will operate when a relief or safety valve opens, or when the tank pressure approaches the relief valve opening pressure.

'The design and application of an electric battery supplied motor driven fan for the purpose of diluting vented or escaping gas from liquid gas tanks and interconnecting piping on a barge or vessel, also to prevent the free escape of an ignitible gas and air mixture, with automatic means sensitive to relief valve action or to the flow of escaping gas to actuate and start the motor driven fan.

'The design and application to a liquid gas tank on a barge, or to a towboat propelling a barge, or to an engine or boiler room of a tanker type vessel, of a gas burning furnace supplied with gas that is vented or discharged from the liquid gas tanks and for which there is no use for propulsion or other purposes; the burner of this furnace being automatically started or shut off by pressure change in the gas supply pipe adjacent to the gas disposal burner, the furnace and burner being such that excess air will be proportioned to limit the temperature of the gaseous products of combustion or flue gases.

'The design and application of an arrangement for providing sufficient solid carbon dioxide (dry ice) to pre-cool a liquefied natural gas tank or holder and purge it of air before filling with liquefied gas, and the application and arrangement of an excess amount of dry ice to automatically purge the tank of gas in case it is emptied of its liquid. Also, an arrangement to contain the dry ice in permeable compartments or containers to prevent its dispersion through the liquid gas by convection currents or other disturbances.

'The arrangement and method of filling an insulated ship or barge chamber or cargo hold, in which metal tanks have been mounted for containing liquefied natural gas at near atmospheric pressure, with sufficient dry ice to pre-cool this chamber or cargo space and purge it of air before any gas enters the tanks mounted in the chamber or the chamber itself. Also the method of filling the chamber and each liquid gas tank or tanks mounted in the chamber with sufficient dry ice to purge both chamber and liquid gas tanks of gas in case they are emptied.

'The design of a combination towing hawser and flexible tubular conduit for connecting a barge or towed hull to a tug boat that would receive its gaseous fuel through the flexible conduit hawser.

'The design and application of an unloading or pumping mechanism for handling liquefied natural gas consisting of a displacement chamber which is alternately filled with liquid, and emptied, by gas being admitted under pressure above the liquid in the displacement chamber thus forcing the liquid out. The return to the main tank of liquefied gas being prevented by a non-return valve which automatically opens to admit liquid when the gas is shut off and the pressure thus reduced in the displacement chamber.

'The design of a venting system for the hold space of a vessel to provide the introduction of sufficient air to control the temperature in the hull, and circulating air providing the heat necessary to maintain the walls of the insulating spaces at any desired temperature.

'The design of such regasification or vaporizing equipment arranged to induce foaming as the concentration of higher boiling point fractions increases for the purpose of forcing the foam into the superheating element where the temperature is higher and where consequently the higher boiling point fractions will be evaporated.

'The design of a liquid natural gas tank or holder with structurally strong inside insulation, in combination with regasification or liquid pumping equipment, that is installed in the tank through an opening in the tank top, the tank top opening being of a relatively small diameter (about 10 per cent of the tank diameter), through which all of the equipment including the inside insulation passes for installation. This design provides a tank with no openings or manholes of any kind below the liquid natural gas surface whether the tank is full or partially empty as well as a tank with a top that is permanently attached.'

'*SAFETY FACTORS*
The design of the liquefied natural gas tanks or chambers with inside insulation for operation at or somewhat above atmospheric pressure provides outstanding safety characteristics some of which are outlined below.

'Inside insulation makes it possible to subject the metal walls of the tank to temperatures that are practically ambient with no atmosphere even though the contents of the tank may be as low as 300° below zero F. Inside insulation that employs material such as balsa or one with similar characteristics, enables the stresses developed by expansion and contraction with changing temperature inside the tank to be taken up by the insulation within the elastic properties of its material.

'The design of the tank with all openings into it through the top adds to the safety in case of accident or inadvertent manipulation, by eliminating the possibility of discharging liquid as would be the case if connections or openings into the tank were below the liquid level.

'The employment of structurally strong inside insulation does not impose on the outer metal walls of the tank any outside collapsing pressure such as that which occurs when loose insulation is used outside.'

But Morrison adds here a warning that some form of equalization of the tank supports must be considered for seagoing ships '. . . where the ship structure deflections may impose unwanted stresses in the tanks . . .'.

That sea-going LNG transport was very much in his mind at this time is evidenced by the rather attractively drawn schematic map of a Mediterranean project.

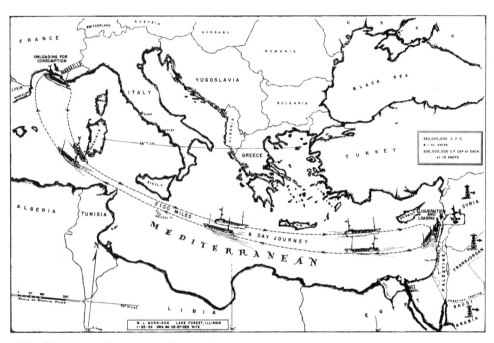

Fig. 2d.  An early ocean going project.

But disappointments lay in store. Difficulties were experienced in obtaining the Classification approvals essential to the project and, much worse, the low temperature tests on the tank insulation, despite the extreme care and ingenuity in its installation, were not successful. Both will be discussed in more detail in later chapters; suffice it to say here that it was felt necessary to revert to the first alternative concept, (a), which in the event proved successful. Despite this setback, however, considerable vital design information on materials and equipment–and also the characteristics of the LNG cargo itself–were developed in this first exploratory phase, added to which, as can be clearly seen, many of the features with which we are now familiar in LNG ship designs were conceived–and patented.

# 3

# Wider Participation and Further Testwork in the USA

The previous chapter discusses the broad lines of approach adopted by the Chicago Stockyard design group in their search for a viable method for barges to carry LNG up the Mississippi, and to establish the cost of so doing. It was soon realized that the concept of water-borne transport had far wider applications than the present scheme; thus any design now developed had to be capable of adaptation to a sea-going ship large enough to exploit the potential markets which undoubtedly existed for the movement of gas from overseas producers to consumers in the USA and Europe. The present 'local' demand for gas in Chicago simply provided the initial stimulus.

For this reason, the approvals of the principal ship Classification Societies were now essential to their endeavours, and the participation of an experienced and competent Naval Architect became indispensible.

Thus, in May 1954, the J. J. Henry Company, at that time a relatively young but forward thinking firm of Consulting Naval Architects in New York, were approached by Willard Morrison with a request

> '. . . to study the technical feasibility of, and draw up tentative proposals for, special ocean-going vessels for the transportation of liquid methane.'

In addition, therefore, to preparing revised drawings for the Liquefied Methane Barges, J. J. Henry immediately set to work to prepare proposals for the conversion of dry cargo vessels of the C-4 type, and T.2 tankers

> '. . . including arrangements, structural solutions and cost estimates and . . . since no precedent existed for this type of vessel, much work was done with various Regulatory Bodies including USCG, Lloyd's Register, American Bureau of Shipping, British Ministry of Transport and Bureau Veritas, to seek their concurrence.'

These efforts culminated in assisting the USCG in codifying tentative regulations, which were later endorsed by the American Petroleum Institute (API).

This early ship design work was shared between J. J. Henry and other companies in the following manner – J. J. Henry developed the vessel arrangements, outlined the cargo handling features and furnished necessary particulars (a) to Gamble Brothers, Louisville, Kentucky, a substantial woodworking company, who designed the hold insulation details and who co-ordinated the testwork, and (b) to Arthur D. Little of Boston who designed the cargo tank details.

In June 1954 a report, which had been prepared by the International Bank of Reconstruction and Development on, amongst other projects, the feasibility of shipping LNG from Kuwait to the UK based on the Constock work was, by a chance meeting on the *Queen Elizabeth*, brought to the notice of the then British Gas Council. British Gas plc, as they are now named, were sufficiently interested to arrange for an immediate visit of one of their senior engineers, Leslie Clark, to evaluate the Constock work. His reports were optimistic.

By the end of 1955 Dr. J. Burns (Chief Engineer) and Mr. Leslie J. Clark (Development Engineer), both of the North Thames Gas Board, were sufficiently impressed by what they had seen of Constock's work to be able to write in their paper of June 1956 'Liquid Methane', now a 'classic':[1]

> 'Natural gas is one of the world's major sources of energy, but is not so readily transportable over long distances as solid and liquid fuels . . . Such gas could be liquefied under atmospheric pressure at temperatures around $-260°$ F and transported in special ships containing insulated tanks and delivered to this country . . . There is at present no precedent . . . for the equipment associated with suitable ocean-going ships . . . some of the design problems are discussed in the paper . . . it is believed that they are capable of satisfactory solutions. The economics of a project are reviewed in broad terms . . . and it is concluded that the scheme has promise . . .'

In fact, it was only three years later, in February 1959, that the first trial shipment was landed at Canvey Island in the Thames Estuary.

Leslie Clark's visit to Pascagoula, Mississippi marked the beginning of a close and constructive association between the British gas industry and LNG ship technology which has continued undiminished through the many phases of its development up to the present day.

British Gas had, from the start, felt that the participation of a major oil company was an essential pre-requisite to the successful commercial development of LNG ship technology, particularly if it was to expand into ocean transportation. Discussions at home bore little fruit despite the fact that the Shell Company, for example, were beginning to look seriously at the problem for reasons already stated. As it so happens, in April 1956 one of Shell's directors remarked, in answer to a question at a conference in London:

> 'In regard to the question of natural gas and its transportation, no one has yet been bold enough to build a ship that will carry the liquid from the places where it is available to the places where it could be consumed, much as the British Government would, I think, like to see that happen. I do not know of anybody, although many are working on it to some extent, who has taken the step of risking the building of a special ship, or the adaptation of an existing ship. It would cost at least £2 million and, when it was built, one could not be sure that any port in the world would allow it to enter and risk the accidents that might conceivably happen, if things went wrong, or that a crew could be persuaded to go on board when collisions or other stresses might release forces that would annihilate them.'

(Today Shell are major participants in the ownership and management of large LNG ships).

Thus, in June 1956, Wood Prince negotiated participation with the Continental Oil Company of Ponca City, Oklahoma, forming Constock Liquid Methane Corporation in which Conoco held 60 per cent of the shares, and the Union Stock Yards of Chicago the remaining 40 per cent.

Other consultants and companies were also retained at this time to advise on various specialized aspects of the work. These included Arthur D. Little (tank design), Oklahoma University (Professor C. Sliepcevich – Chemical Engineer), Gamble Brothers, Louisville (wood working) and Purdue University (Dr. Chenea). With this considerably expanded team, the injection of additional funds and the continued encouragement from the British Gas Council, Constock now set out to develop a design based upon an independent, self-supporting tank, built of a material suitable for operating at LNG temperature ($-258°$F) and located in the insulated hold of a ship – the insulation being designed to protect the hull structure in the event of leakage from the tank.

Although a basic cylindrical shape was not altogether discounted, efforts were concentrated on developing a flat-sided, box-shaped design which would better utilize the underdeck volume of the ship.

The state of the art at the time is best described by summarizing the major design problems which needed to be solved by test work – or other means – before a workable proposal could be produced.

**Tank materials:** Stainless steel was known to have good low temperature properties but was much too expensive and 9 per cent nickel steel was still not sufficiently well proven. Aluminium therefore seemed to be the only possible material to use, but reliable information on its mechanical or physical properties at low temperatures was still not altogether satisfactory, particularly at the heavier gauges; furthermore, welding – to achieve the required standard of mechanical properties (the highest possible) – was relatively undeveloped and quite difficult to achieve. Fatigue properties of neither parent metal nor weld at $-260°$F were fully established; even its coefficient of thermal expansion at LNG temperature had not been fully confirmed at this time, so that stress levels under normal operating conditions were difficult to define with accuracy. A test programme was therefore set up to rectify this lack of information.

As a member of Lloyd's Register reported during a visit to the USA at this period:

> 'A fabricated aluminium angle section, welded at the corner, was immersed in liquid air ($-290°$F), removed and subjected to hammer and bend tests. The angle was flattened by hammering and then doubled on itself without fracture and appeared to have lost none of its ductility at this temperature.'

The same surveyor went on to state:

> '. . . . it may be of interest to note . . . that extensive tests are being carried out on a 9 per cent nickel (non-chromium) steel having excellent low temperature properties and extremely high tensile properties at room temperature. This steel is said to be three to five times more economical than stainless steel and to give similar low temperature properties.'

**Insulation materials:** Attention was focused principally on balsa-wood as this seemed to be the only material able to combine insulating properties with the ability to contain methane and thus act as a secondary barrier. Sufficient tests had already been carried out to satisfy everyone concerned that balsa

- – was unaffected by contact with hydrocarbons;
- – could be worked to close tolerances by existing woodworking techniques (provided the tools were sharp);
- – could be bonded to itself with adhesives able to withstand liquid methane temperatures;
- – could withstand sufficient compression loads to support the tanks;
- – had sufficient elasticity at low temperatures;
- – had a relatively low coefficient of thermal expansion over the whole operating temperature range;
- – was relatively unaffected by low temperatures even if moisture was present;
- – was commercially available at a supportable price.

It only remained to devise a means of assembling it in such a manner that it would contain liquid methane in an emergency for a sufficiently long period – and that it would continue to perform this function over a period of years when installed in the flexible structure of a ship.

**Tank design criteria:** There were a number of facts which required to be established before the design for the tank and perhaps more important, supporting arrangements, could be undertaken.

Ship accelerations: No quantitative information of any kind was available at this time. 'Informed' opinion however agreed that a reasonable assumption for design purposes would be the combination of a 30° roll (half amplitude) having a ten second period, *with* a 4° pitch at eight second period *plus* a horizontal and vertical vibration of $\frac{1}{2}$ inch amplitude at about 120 rpm. Since '. . . it is extremely unlikely that all of these motions would occur simultaneously . . .' and the fact that slamming and heaving were being ignored and that in any case '. . . the effect of slamming is virtually incalculable . . .' these assumptions seemed, on the whole, to be justified.

Cargo density: Liquid methane density is about 0.42, but it was considered prudent to design for liquid propane at 0.6.

Tank support: This aspect of the design presented most of the headaches – and produced many of the early patents! Somehow arrangements had to be made for the tank

(a) to expand and contract freely both horizontally and vertically;

(b) to remain in position when the ship rolled, pitched and vibrated;
(c) to absorb normal ship hull deflections – in themselves unquantifiable at this time;
(d) to absorb the severe temperature gradients which were expected to occur during filling and emptying

**Design concept:** To meet this combination of requirements a design was developed which, in the event, became the basic Constock/Conch tank design. It was essentially a flat-sided, horizontally stiffened tank with a bottom chamfer to accommodate the normal ship bilge construction. A test tank of about 75 m³ capacity incorporating all the basic design features was built at the Arthur D. Little works in Cambridge, Mass, (near Boston) and fully instrumented. Balsa insulation was fitted to the bottom and three sides

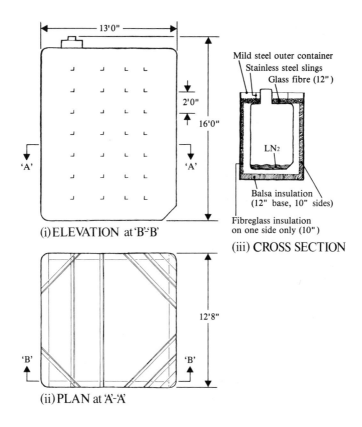

Fig. 3a. A. D. LITTLE–LNG Test Tank.

of the inside of the mild steel outer casing, which represented the ship's hull; 12 inch glass fibre was laid on the top and attached to the remaining side. The tank was slung on stainless steel cables to simulate the top keying arrangement proposed at that time. After being 'spray cooled with liquid nitrogen to avoid stress concentrations', it was filled to a depth of two feet with liquid nitrogen ($LN_2$).

Vibration of the ship was simulated by an electric motor connected to a 40 inch rotating arm with a 25 lb weight at the end; 'the effect was very similar to an extremely bad case of ship vibration at about 95 cycles/minute'. At the same time pressure pulsations, achieved by shutting in and releasing the relief valve, were imposed in an attempt to simulate ship movements at sea.

To simulate slamming of the ship the end of the tank, supported by a crane, was dropped to the ground through a distance of about $2\frac{1}{2}$ inches. The end of the tank was then lifted about 3 inches and again dropped. 'The impact was considerable in both cases as the combined weight of the tanks and fluid was about 75 tons, but no damage appeared to have been sustained.'

To simulate a full tank, the complete tank was laid on its side thus immersing the entire side structure in $LN_2$. Finally $LN_2$ was poured into the insulation space to check that the balsa was capable of containing the liquid for a sufficient period. In fact, 12 inches of $LN_2$ boiled off in six hours but there was no indication of liquid leakage during this time. Thus balsa was established as a satisfactory secondary barrier.

Not only was valuable design data on stress levels in the structure, boil off and temperature gradients obtained from this series of tests, but their substantial scale provided the many organizations who witnessed them with a considerable degree of confidence in the safety and viability of the eventual outcome.

In addition to the above-mentioned test programme on the containment system as a whole, detailed studies were being carried out on the component parts of the system; particular attention was paid to balsa panel design and thermal gradients through such panels, the permeability of plywood and balsa and its resistance to fire; the mechanical properties and welding of various aluminium alloys were also being investigated in depth.

The Boston tests provided a wealth of design data and also served to obtain approval in principle from Lloyd's, ABS and the USCG for the proposals which had by now been drawn up by J. J. Henry for the conversion of a T.2 tanker, in which the LNG would be carried in eighteen aluminium cargo tanks, six tanks being fitted in each of three insulated holds; the tanks were to be constructed of Kaiser's newly developed 5083-0 aluminium alloy, or equivalent.

Plate 1. The outer tank casing and protective cover.   Plate 2.   Installing the Balsa panels.

Plate 3.  Lowering the Aluminium tank into position.

29

Plate 4.  General view – cold test in progress.

The first LNG test tank facility at Boston, Mass.

'The (tank) design is based on a 0.6 SG liquid with a head equal to the full depth of tank and trunk plus dynamic forces plus 3 psi . . . rolling an assumed 30°. This design allows the tanks to be tested in a static condition full of water, plus 4 feet head above trunk top. Design stress in the plate is to be 12,000 psi maximum, and the design stress of the cross bracing is to be 12,000 psi maximum. [This latter figure was subsequently reduced to 10,000 psi.]'

In parallel with all this work on the design of the ship's containment system, a long term study and demonstration had been mounted at Lake Charles, Louisiana, in collaboration with the API and US Bureau of Mines, to investigate the burning characteristics of a large (20 foot × 20 foot) burning pool of LNG in comparison with a similar pool of gasoline. Fire extinguishing tests, using both water, foams and a variety of dry powders were also carried out. The result of this extensive test programme, of which filmed records were made and are still available, was to satisfy the Authorities including, subsequently, the British Government, that liquefied methane presented no special hazard provided proper precautions were taken, and proper materials were used in handling and storing it. (See colour plate following page 134.)

The British Government's 'Petroleum (Liquid Methane) Order, 1957, No. 859' classified liquid methane as a petroleum spirit . . . no more dangerous than other volatile petroleum derivatives.

---

[1] 'Liquid Methane' by Dr J. Burns and L. J. Clark. Institution of Gas Engineers, London, 1956.

S T A T U T O R Y     I N S T R U M E N T S

## 1957 No. 859

## PETROLEUM

### The Petroleum (Liquid Methane) Order, 1957

| | | |
|---|---|---|
| *Made* - - - | 17*th May*, 1957 | |
| *Coming into Operation* | 1*st June*, 1957 | |

At the Court at Buckingham Palace, the 17th day of May, 1957

Present,

The Queen's Most Excellent Majesty in Council

Her Majesty, by virtue of the powers conferred on Her by section nineteen of the Petroleum (Consolidation) Act, 1928(a), is pleased, by and with the advice of Her Privy Council, to order, and it is hereby ordered, as follows:—

**1.**—(1) The provisions of the Petroleum (Consolidation) Act, 1928, except the provisions specified in the Schedule to this Order, shall apply to liquid methane subject to the modifications specified in paragraph (2) of this Article.

(2) (*a*) Section five (which contains provisions as to the labelling of vessels containing petroleum-spirit) shall have effect as if, in the words required by that section to be shown on the label referred to therein, the words "Liquid Methane" were substituted for the words "Petroleum-Spirit".

(*b*) Subsections (1) and (2) of section seven (which requires byelaws to be made for harbours as to the loading of ships with petroleum-spirit and generally as to the precautions to be observed with respect to ships carrying petroleum-spirit whilst in the harbour) shall not apply except in relation to a harbour in which there is an installation specially constructed for landing liquid methane from ships or for loading liquid methane onto ships and in relation to the harbour authority having jurisdiction therein.

**2.** This Order may be cited as the Petroleum (Liquid Methane) Order, 1957, and shall come into operation on the first day of June, 1957.

*W. G. Agnew.*

SCHEDULE             Article 1 (1)

PROVISIONS OF THE PETROLEUM (CONSOLIDATION) ACT, 1928, NOT APPLIED TO LIQUID METHANE

The proviso to subsection (1) of section one.
Section ten.
Section seventeen.
Section twenty.

(a) 18 & 19 Geo. 5. c. 32.

Fig. 3b. The U.K. Government's Liquid Methane Order No. 859–1957.

# 4
# Activities Outside the USA

Of course, everyone knows that Constock led the world in LNG ship technology in the 1950s; and, as Conch, held on to this lead well into the 1960s. Their position seemed unassailable. But a cursory examination of the state of the art in the mid-1950s shows that a surprisingly high level of engineering study had been directed to this new technology, with its almost magical opportunities, in other parts of the world. Complex – yes; insoluble – definitely not; if the Americans could do it, then Europe would not be far behind.

As already briefly mentioned in an earlier chapter, the Shell Group had already started work by 1955 – drawing upon the considerable resources available within the Group, both in London and The Hague and their respective research facilities.

Also in London, still, after all, the bastion of shipping enterprise, Consulting Naval Architects Burness, Corlett & Partners had been commissioned to carry out economic and design studies into the carriage of liquefied natural gas.

In Norway – also the home of many redoubtable shipowners – the Lorentzens had patented a spherical concept in 1955 and were now (in 1956) offering such a ship of 15,000 tons methane deadweight, with classification approval in principle, albeit with somewhat insubstantial design back-up.

In Germany and Italy, shipyards and shipowners were mainly concerned with the problems associated with the much 'warmer' LPG.

In Japan, by contrast, shipyards were all totally committed to the development of VLCC production on a grand scale; no attempts were made to develop the high technology ship until well into the 1960s.

The only other outward visible sign of serious study into LNG ships was in the USA itself where a John Crecca, of Lehigh University, produced a thesis on LNG transportation, which included a T.2 conversion; his design was clearly influenced by current activities elsewhere in his own country and based on little or no test work of any kind. The design certainly contained some innovative thinking; it created worldwide interest at the time of its publication but, with 4,000 m³ LNG capacity in eight large and two small stainless steel rectangular tanks which were internally insulated, each with a balsa wood swash bulkhead, it was not followed through and did not become a serious contender.

Let us examine in a little more detail these other contemporary efforts outside the USA.

## THE SHELL STUDIES

Shell had initiated, by 1955, conceptual ship design studies and a materials survey with particular emphasis on finding an insulating material which could be attached to the inside of a metal tank.

The ship design concept was allocated to the Marine Department in London; the search for insulating materials was co-ordinated in The Hague and Amsterdam. This allocation of responsibilities was not without its difficult moments, however, because the Engineering Group in The Hague

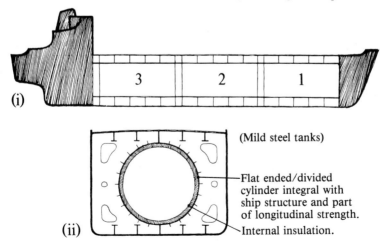

(Mild steel tanks)

Flat ended/divided cylinder integral with ship structure and part of longitudinal strength.

Internal insulation.

Fig. 4a. HAGUE solution.

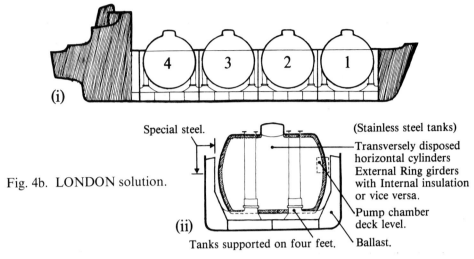

(i)

Special steel.

Fig. 4b. LONDON solution.

(ii)

Tanks supported on four feet.

(Stainless steel tanks)

Transversely disposed horizontal cylinders External Ring girders with Internal insulation or vice versa.

Pump chamber deck level.

Ballast.

was convinced that the LNG tanks should be cylindrical in form, internally insulated and an integral part of the ship's structure (fig. 4a), whereas London was equally convinced that it was essential that the LNG tanks be completely separated from the ship's hull and its associated stresses and strains (fig. 4b). Both envisaged internal insulation, but with typical British caution, the London solution could much more easily be adapted to an external insulation system if this was forced upon them.

The search for an insulating material to meet the design requirements ended with the same solution as that adopted by researchers in the USA – no known material, other than balsa wood, was found which could withstand the basic test of immersion in liquid methane for a given period of time, removal and warm up on the laboratory bench, without varying degrees of disintegration during warm up. In other words, no satisfactory internal liquid-tight insulation could be found. Balsa itself was permeable to an extent but all hope was pinned on the 'back pressure principle', which meant that as the cold liquid penetrated the balsa it would vaporize and, being prevented from escaping through the warm side, would develop sufficient pressure to prevent further ingress of liquid.

Shell therefore pressed ahead with a test programme in which an existing aluminium LPG cylinder of approximately 10 feet diameter and 20 feet long would be internally insulated with balsa, then filled – initially with refrigerated propane, subsequently with LNG. The code name for the test rig was ALICE (aluminium/ice).

The balsa was meticulously installed in cross-laminated sections – but not under compression – by a UK importer of balsa wood for model kits, with a joinery works on the Sussex coast – Solarbo Ltd. – and duly instrumented and filled with refrigerated propane; however the balsa contracted, the joints opened and widespread frosting of the outer shell was observed. This was clearly not the solution. However, before an alternative could be developed, all further research was abruptly terminated following the outbreak of the first Suez War. Shell's tentative plans for a 250 foot, 750 ton capacity 'pilot tanker' were also shelved at this time.

## THE BURNESS, CORLETT WORK

This firm of Consulting Naval Architects, based in London and Basingstoke, had been commissioned by two groups to work on LNG; first by Westinform shipbrokers, with a request to study the economics of LNG ship transport, and secondly, a much more substantial contract from Wm. Cory & Son in 1958 to develop a design for a ship of around 14,000 tons methane carrying capacity. In common with current thinking at the time, Burness, Corlett's studies were based on a cylindrical form of containment, since this basic shape offered the greatest degree of freedom from stress concentrations, combined with economy in the use of expensive cryogenic material. Their studies were concentrated on two basic arrangements:

(1) multi-horizontal cylinders supported by saddles, located both below and above deck to obtain the requisite carrying capacity. This design was proposed and argued by Sir Denistoun Burney, a

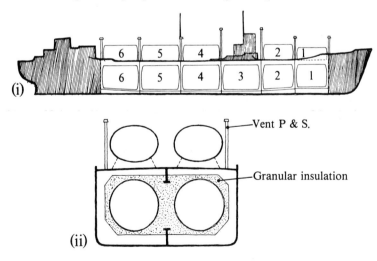

Fig. 4c. DENISTOUN-BURNEY solution.

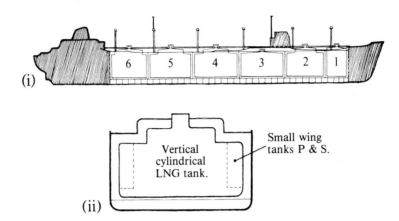

(i)

(ii)

Small wing tanks P & S.

Vertical cylindrical LNG tank.

Fig. 4d. BURNESS, CORLETT solution.

British inventor of some note and associated with Cory's at that time.

(2) Vertical cylinders of larger diameter keyed into the ship at their axes–this was their own preferred solution.

Both solutions were based on an external insulation system of some kind.

Areas of particular concern to Burness, Corlett were the insulation: how it should be attached; how to compensate for the expansion and contraction of the tanks (this presented problems whether it were to be fibrous glass insulation, or if a powder form of insulation were to be used, how to prevent its 'compaction' after several thermal cycles of the tanks, with consequential external overpressures).

One of the national UK research laboratories was allocated a contract to investigate the whole question of insulation design–meanwhile Burness, Corlett's own ship concept had been selected as the preferred arrangement upon which all further study should be concentrated.

In fact their work was published in considerable detail some years later in a paper presented to the Institution of Naval Architects[1] as a valuable contribution to the art at a critical stage in its development. Given a greater allocation of funds, backed up by a trial prototype, who knows how far this design, which showed great prescience in many respects, would have developed? Perhaps, in hindsight, the same distance as the *Jules Verne* (now renamed *Cinderella*), a very similar concept, which has remained unique to this day.

## DR. LORENTZEN'S SPHERE

As already mentioned, the Norwegian ship-owner, Oivind Lorentzen of Oslo, had designed and patented a methane tanker by mid-1955. Dr Lorentzen offered the design in its early form to a number of major shipowners and also to the British Gas Council, having, it must be assumed, reached the stage where additional funds were essential for the final phase of testing and evaluation. For good measure he was able to claim Det norske Veritas approval in principle.

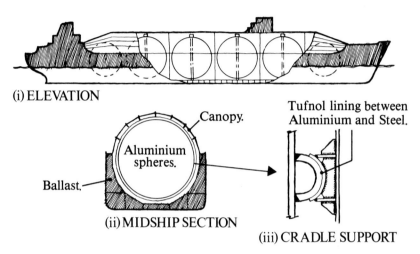

(i) ELEVATION

(ii) MIDSHIP SECTION

(iii) CRADLE SUPPORT

Fig. 4e. LORENTZEN'S concept for 17,000 ton Methane tanker.

The design was interesting in that it comprised six aluminium spheres, four being of 24 m diameter and two of 20 m diameter, each supported by a circumscribing annular ring around a line of longitude (in preference to the equator) incorporating means for passing oil or other suitable fluid through the annular duct thereby maintaining the temperature at the outermost surface of the aluminium ring at or about the ambient. In this way it was expected that the steel supporting ring which was integral with the ship's structure would not be subject to low temperature conditions. In addition to the external insulation, it was proposed to fit a thin layer of internal insulation to relieve the metal walls from shock effects when the tanks were being filled.

The Lorentzen design, sadly, joined the ranks of good ideas and paper studies, and it was not until over ten years later that a spherical design was reintroduced to the LNG scene.

## ACTIVITIES IN FRANCE

It was only one or two years later, in 1957, that the French Government, activated by the discovery of the vast gas reserves in Hassi R'Mel, Algeria – at that time a French 'possession', requested the Worms Group to study ways and means of exploiting these reserves – and it was M. Worms himself who initiated investigations into LNG marine transportation.

Meanwhile the major French shipyards were carrying out their own studies into potential materials and techniques for use at temperatures of $-165°C$. Each yard produced its own individual LNG containment solution, three of which were later evaluated on the seagoing prototype *Beauvais* to which Chapter 7 is devoted.

The early mid-1950s was a period when many organizations and individuals, being convinced that LNG could – and must soon – be transported by sea, were working towards

- a definition of the design parameters;
- locating suitable materials and establishing their properties within a preferred design framework;
- establishing the cost of the ships when built and estimating their optimum size.

All these groups were tantalized by 'reliable' reports that the senior Naval Architect of the J. J. Henry Company in New York, now well integrated into the Constock team, was working an 18-hour day with a pile of blank cheques on his desk – his sole terms of reference being to 'make damn sure that you come up with a workable solution'. (This report has subsequently been checked and found to be totally untrue in common with many of the 'reliable reports' that have been heard, and still are, in the LNG tanker field since that time.)

Nevertheless, rapid progress was clearly being made in the USA, but cloaked by an impenetrable veil of secrecy. Nothing useful, nothing at all, could be gleaned by those outside to help solve their problems except, perhaps, that balsa was in some way being used successfully for the insulation system.

In effect, it was in the matter of the wide range of supporting test work that Constock scored so heavily in these early years.

Competent designers could produce, and were producing at this time, a number of well reasoned and practical solutions to the many technical problems involved, but, in the final analysis, none could state with any real confidence whether their solutions would function reliably and safely, since no relevant experience existed. Large scale test work was an absolutely vital ingredient – and it was cripplingly expensive.

[1] 'Methane transportation by sea' by Dr E. C. B. Corlett and Dr J. F. Leathard. RINA, London, March 1960.

# 5
# The Role of
# the Classification Societies

The Classification Societies' raison d'être is to maintain a standard; their recommendations are based very largely on their uniquely comprehensive records of ship performance data, but also on their capacity to analyse from first principles and to integrate this capacity with performance in service. There may be many people involved in design, construction and operation of ships whose experience of Classification Societies and other regulatory agencies is one of submitting drawings or proposals to a disembodied office and having them returned several months later liberally annotated in red—frequently to the detriment of their schedules and budget; LNG ship designers are not among them.

In embarking upon a new technology it is essential to understand that where no previous experience exists, no single organization can be expected to provide instant answers or solutions to the many questions which arise on a day to day basis. There are no 'rule books' to which one can refer.

Thus when the American Bureau of Shipping received a letter from Ingalls Shipyard, Pascagoula, Mississippi on 2nd March 1953, with the customary three copies of a drawing of a barge designed to carry liquefied methane for river service, the Bureau replied, and with remarkable promptitude–nine days later, that although

'. . . based on a preliminary analysis the scantlings are considered acceptable for River Service . . . with regard to the intended cargo we have no experience with the carriage of liquid methane. Accordingly before granting approval it would be necessary that we be furnished with full information of the means of loading, unloading and all pertinent characteristics of the gas or liquid and the state in which it is carried, i.e. temperature, toxic qualities, flash point, etc. It is proposed that the Coast Guard be contacted with regard to compliance with their regulations for this vessel. We note that a pump room and heat exchanger equipment compartments are indicated in spaces adjacent to the tanks and advise that under our requirements for vessels carrying cargoes having a flash point below 150° F vapor igniting equipment would not be permitted in spaces adjacent to the cargo.'

Subsequent correspondence seems to indicate that, not surprisingly, it was many months before either the 'prospective owner', the shipyard, the Bureau or the USCG came to agreement on this project which was, after all, to quote Ingalls

'. . . of an experimental nature and . . . operational details will have to be worked out as the work progresses.'

Reading between the lines of the extracts from this early correspondence it does seem that the shipyard adopted a somewhat cavalier, if not naïve, approach to this totally new technology.

LNG struck Lloyd's a year or two later when the British Gas Corporation requested their participation to assist in their evaluation studies; Bureau Veritas were approached a little time after that; each reacted in the same way, since it was, of course, as new to them all. The real problem lay in the total absence of both relevant experience and basic design data, either in the ship configuration, suitable materials of construction, or the character of the cargo itself.

Each organization, however, realizing that the problem was unlikely to 'go away', set out to establish some reliable data on which to base their evaluation of these designs, which were soon being submitted by a number of organizations and in varying stages of development.

For those directly involved in the search for a solution–for which Classification Society approval was a sine qua non–the working relationship which developed (and still exists) between them was a memorable experience; both were seeking the same ultimate goal–a design which was not only inherently safe, but reliable and economically viable.

The Classification Societies made available, to all serious enquirers, the full depth of their experience, technical expertise and judgement. Realistic guidelines were jointly produced; information was exchanged as far as confidentiality would allow; every Society was meticulous in its observation of client confidentiality. Indeed on one occasion, the author, after discussing a new design development with the then Chief Surveyor of Bureau Veritas and timidly asking his assurance that the subject of the discussion should be kept strictly confidential, M. Blanc drew himself stiffly upright and replied, 'Monsieur Ffooks, if you had asked me that question fifty years ago I would have drawn my sword!' Confidentiality in the 1950s and 1960s was a real problem for the Classification Societies for each design group was quite desperate to know what the other was doing.

A further nightmare situation for the Classification Societies was that of falling into the trap of recommending procedures or solutions which automatically led to one client infringing the patents of others. It was all too simple jointly to work out a general solution and to base classification guidelines upon it, only to find that it had promptly been patented and could, thereafter, no longer be recommended as a 'guideline'. This became a serious problem while basic criteria were being established, both for the original and later designs. Who could imagine, for example, that a double hull ship could be patented? The *Great Eastern* – designed by Isambard Kingdom Brunel in 1858, exactly 100 years earlier – was so fitted, but Constock obtained world-wide patent coverage for it, with a few associated features besides. An unfair generalization, true, but this was the atmosphere in which the technology developed during the period and, indeed, well into the early sixties.

Although the ABS were the first to evaluate the basic problems of LNG containment *per se*, Lloyd's seemed to be the first to grapple with the sea-going aspects, as part of their work for British Gas who had now commenced serious discussions with Constock on the possibility of converting a ship for ocean-going service. The intention was to carry a number of trial cargoes across the Atlantic to a proposed terminal at Canvey Island, and subsequently to enter into negotiations for a long term gas importation project. If they, British Gas, were to share the cost of such an experiment, Lloyd's approval of the technology would be mandatory.

It was thus during an early visit of inspection to the Boston test facility paid by Ian McCallum, one of Lloyd's Senior Surveyors, that agreement was reached on a definition of ship accelerations; based on little more than common sense and a longhand calculation. Looking back at this period it is really quite surprising that, in the whole history of Naval Architecture, one should have had to wait until 1957 – and LNG – for the first attempt to produce this piece of very basic ship design data.

Meanwhile, on both sides of the Atlantic investigatory work was being initiated into confirming that the 5,000 series aluminium alloys were, indeed, suitable for marine service.

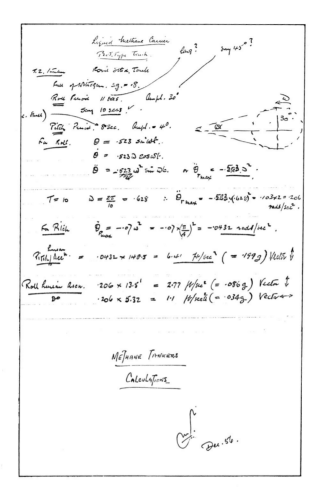

Fig. 5. A sheet from Ian McCallum's note book–1956.

Particular attention was paid to its ductility and notch toughness; and since the Charpy Vee notch test could not be used on a comparative basis with steels, the US Navy tear test was adopted as a standard: the objective was to satisfy all parties that under no foreseeable conditions could brittle fracture occur, particularly in the higher thicknesses which would be required for commercial ship tanks. Lloyd's carried out a series of tests in their own laboratories at Crawley;[1] ABS participated in parallel US industry sponsored work. Bureau Veritas and Det norske Veritas were similarly engaged in their respective countries.

42

As far as insulation was concerned, all the Classification Societies tended to rely on the system tests carried out by the designers themselves, the test programmes for which were for the most part approved beforehand and the final results witnessed.

The exploratory period was, therefore, one of a close working collaboration between all parties concerned and although each of the Societies had drawn up their own internal guidelines during the mid-fifties, which were 'unofficially' available, published 'recommendations' were not officially issued until much later, ie early 1958–USCG;* July 1960–Det norske Veritas; January 1961–Lloyd's; and September 1962–Bureau Veritas.

These guidelines were all, to a very large extent, based on the rationale developed in the USA but all contained sufficient flexibility to allow consideration and approval of alternative arrangements–including membrane, or integrated, designs. A secondary barrier, to provide 'temporary' containment of leakage from the primary tanks for an unspecified period, was required by all Societies; however DNV took a more sanguine approach to the risk of primary tank leakage, and were at the same time perhaps a little more specific; they required:

> 'For ships designed for transport of liquefied gas with boiling point at atmospheric pressure below $-50°$ C, the average heat transfer coefficient mentioned in (a) should be reduced in relation to the lowering of the boiling point below $-50°$ C. The insulating layer should also be carried up the sides of the holds to a height of 0.2 B above the inner bottom and covered with a watertight layer of aluminium or some other satisfactory material, where B = moulded breadth.'

(a) required that the average heat transfer coefficient of the insulating layer should not exceed 2.0 kcal/m$^3$ h$°$C.

Classification requirements in regard to the important matter of ship accelerations, fundamental to the design of the containment, remained far from uniform for a number of years, with quite wide disparities in the required values for roll and pitch and their phasing, and also in the reduction allowed, if any, for increase in ship size. Although this simply reflected the state of current knowledge of ship behaviour in a seaway, it remained a great inconvenience to designers until rationalization finally appeared in Chapter 4 of the IMO Code – this being, as it so transpired, one of the three chapters produced exclusively by the Classification Societies themselves through IACS (International Association of Classification Societies).

* The USCG tentative standards, as revised by the API Special Committee on the Transportation of LNG by Water, 22 August 1956 and unofficially available at that time, are reproduced in full in Appendix 3.

Surge forces due to pitching, deceleration due to collision, buoyancy forces due to flooding, hull steel grades, relief valve capacity in relation to fire exposure, were among many secondary design requirements upon which agreement took many years to reach. The Thomas/Schwendtner paper, 'LNG Carriers–the State of the Art', presented to the Society of Naval Architects and Marine Engineers in New York in November 1971[2] provided a detailed scenario of this 'situation' which has now, happily, almost disappeared.

The title of this chapter, 'The Role of the Classification Societies', was chosen with some deliberation. This role has, of course, remained unchanged since their formation–to establish and maintain an acceptable standard of ship construction and maintenance. In establishing a standard for LNG ships they had to rely heavily on their evaluation of the work and experience of the designers themselves, having no in-house records to which to refer and relatively modest conceptual design and test facilities; today the position is almost completely reversed, for very few, if any, organizations can match the breadth of experience and design capability possessed by any one of the Classification Societies. Thus although the conceptual initiative may still originate from without, the major in-depth design is or can be, if necessary, carried out within their own resources.

While there is no doubt about the contributions made to the development of LNG technology by the Headquarters staff of the Classifcation Societies, no recommendations or rules can be translated into a viable ship without the contributions of their outside staff; the early days of LNG found many dedicated outside surveyors–some now in deservedly senior positions; theirs was, and still is, a difficult task in which exceptional, on-the-ground judgement was required and, being given, was much appreciated by all concerned.

[1] 'Low temperature properties of some aluminium alloys' by J. E. Tomlinson, D. R. Jackson, R. J. Durham, J. Sawkill, D. James and R. E. Lismer. A seminar, Aluminium Development Association (now Aluminium Federation), England. Publication No RP77, 1958.
[2] 'LNG carriers: the current state of the art' by W. D. Thomas and A. H. Schwendtner. SNAME, Vol 79, 1971.

# M.V. 'METHANE PIONEER'

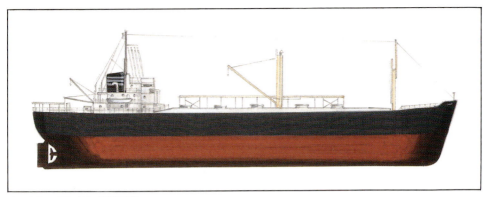

(i) GENERAL VIEW

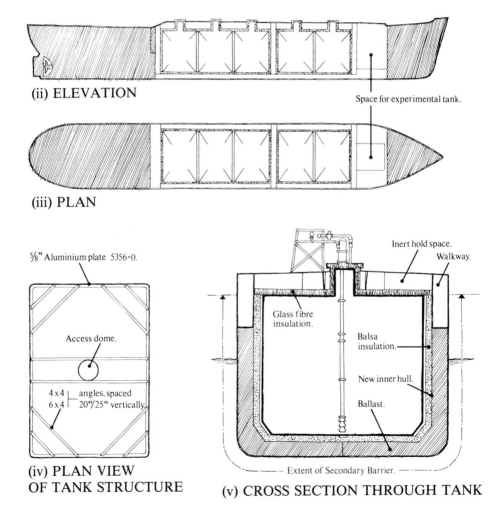

(ii) ELEVATION

Space for experimental tank.

(iii) PLAN

⅝" Aluminium plate 5356·0.

Access dome.

4 x 4 ⌐ angles, spaced
6 x 4 ⌐ 20"/25" vertically.

(iv) PLAN VIEW
OF TANK STRUCTURE

Inert hold space.

Walkway.

Glass fibre
insulation.

Balsa
insulation.

New inner hull.

Ballast.

Extent of Secondary Barrier.

(v) CROSS SECTION THROUGH TANK

# 6
# The First Prototype – *Methane Pioneer*

The decision to proceed with the construction of a prototype LNG ship was finally reached in 1957 when the British Gas Council agreed to participate on a 50/50 basis in such a scheme. It was considered by both parties that no amount of calculations and small scale tests, essential as these undoubtedly were, could provide the information and the confidence needed for the final proof that transocean shipment of LNG was a safe, practical and economic proposition; a proposition, furthermore, which could be confidently recommended to the many authorities and organizations which would be involved in an LNG import scheme, not the least important of which was Her Majesty's Government in the UK.

Having taken the decision in principle to go ahead, the next question was, what size? And at what cost? The basic philosophy was that the ship should be

(a) large enough to provide realistic design data on which to base a full scale commercial project;

(b) the initial cost should be kept as low as practicable consistent with (a) above; however it was foreseen that, if the vessel could be used for commercial LPG trading after her evaluation trials, this could in time recover all or part of the initial investment;

(c) the capacity should be sufficient adequately to prove the land storage facilities at the Canvey Island reception terminal.

45

J. J. Henry were allocated the task of examining the feasibility of several alternative schemes–the two most favoured being the conversion of a T.2 tanker and the conversion of a C1-M-AV1 dry cargo ship–both US wartime built ship types and readily available on the second-hand (used) ship market. In the event the latter was selected.

Part of the British Gas Council bargain was that the ship should be classed by Lloyd's Register–the first indication of the political pressures that were, though much later, to dominate the LNG market, but in the event the conversion was carried out under the dual classification of Lloyd's and ABS.

Under the heads of agreement with Constock:

'(1) The Gas Council jointly with Constock should buy and convert a second-hand tanker to carry liquid methane, at a cost, including conversion, to the Council of about £700,000, and a further £300,000 for an unloading installation at Canvey

(2) The running of this ship on an experimental basis would be on a 50/50 basis.

(3) If the experiment was a success it was the intention of both parties, but without commitment, to build and operate a commercial tanker.

(4) A joint company should be set up to operate the venture.'

**THE CONVERSION**

The specification drawn up by J. J. Henry during the latter half of 1957 and formalized in February 1958, for the conversion of cargo motor vessel m.v. *Normarti* into a vessel to carry 5,000 m³ of liquefied natural gas in bulk at atmospheric pressure, required that:

– the original poop deck be extended forward to the forecastle deck to form a new upper deck;

– longitudinal wing bulkheads (with bottom chamfer) be fitted; wing tanks, so formed, to be combined with double bottom tanks for ballasting;

– the wing spaces between the original and new upper decks will form access walkways;

– cargo will be carried in five aluminium cargo tanks which will be inserted in the new holds; two forward and three aft;

– the holds will be completely insulated;

– the conversion is to be made in accordance with the requirements and subject to the approval of the American Bureau of Shipping and

Lloyd's Register of Shipping . . . the US Coast Guard will approve plans and inspect the conversion.

More specifically, the main component parts of the ship were designed and fitted as follows:

**The Hull:** No requirements for special steel grades were imposed; however:

'Due to the fact that prefabricated insulation will be fitted against the longitudinal and transverse bulkheads in the cargo holds, every effort shall be made to provide steel work which is smooth . . . and exactly to dimensions given on plans . . . the distance "off centreline" within a tolerance of $\pm\frac{1}{2}$ inch . . . the distance between bulkheads within a tolerance of $\pm\frac{1}{2}$ inch.

Steelwork of inside of new cargo holds, except for the underside of the deck, to be sand blasted and left unpainted.

The [wing] walkways shall be provided with a heating system; and a salt water spray system shall be installed in the new deep tanks [cofferdams], wing double bottom tanks and walkway – supported with salt water from the fire main.'

It was considered essential that means should be provided for applying heat to the inner hull as a means of combating 'cold spots' in the event of local insulation failures.

**Insulation** consisted of:

'. . . prefabricated sandwich-type panels of balsa wood faced with maple and oak plywood. The panels are to be attached to bulkheads and tank top by means of Nelson studs and are to be installed with a system of seals and expansion fittings which will prevent convection movement of gases in and around the insulation and provide a liquid-tight interior surface.'

In the event the panels were prefabricated and installed by Gamble Brothers of Louisville, Kentucky, who were also responsible for much of the early design and test work on the system.

Prior to installing the balsa panels, the steel bulkheads had been coated with a trowelled on mastic to provide a smooth surface to which the panels were subsequently directly applied.

During installation of the panels the humidity in the holds was carefully maintained at 30 per cent or less to prevent moisture pickup by the wood; furthermore the temperature level in these spaces was maintained at 80° F or more to provide for adequate curing of the adhesives used.

The tops of the tanks were insulated by eight layers of 2 inch glass fibre batts, with staggered joints.

Figure 6b shows how the insulation was arranged and fitted in the ship.

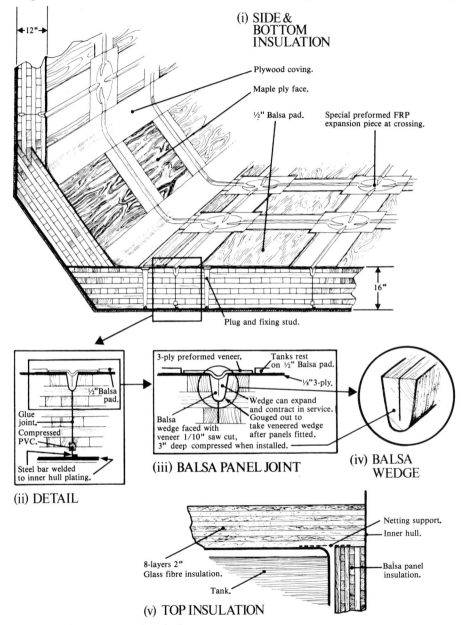

(i) SIDE & BOTTOM INSULATION

Plywood coving.

Maple ply face.

½" Balsa pad.

Special preformed FRP expansion piece at crossing.

16"

Plug and fixing stud.

(ii) DETAIL

½"Balsa pad.

Glue joint.

Compressed PVC.

Steel bar welded to inner hull plating.

(iii) BALSA PANEL JOINT

3-ply preformed veneer.

Tanks rest on ½" Balsa pad.

⅛"3-ply.

Balsa wedge faced with veneer 1/10" saw cut, 3" deep compressed when installed.

Wedge can expand and contract in service. Gouged out to take veneered wedge after panels fitted.

(iv) BALSA WEDGE

(v) TOP INSULATION

8-layers 2" Glass fibre insulation.

Tank.

Netting support.

Inner hull.

Balsa panel insulation.

Fig. 6b. *Methane Pioneer* insulation system.

**Cargo Tanks:** These were designed by the Arthur D. Little company, with the assistance of consultants from Oklahoma University: the arrangement of internal structure conformed very closely to that tested in Boston, as described and illustrated in an earlier chapter. Design stress levels were modest at 10,000 psi for the plating and 8,000 psi for the extruded stiffeners. Ships motions were assumed to be the addition of

| | | |
|---|---|---|
| Pitch | – | 6° above and below horizontal |
| | | 8 second period |
| Roll | – | 30° each side of the vertical |
| | | 14 second period |
| Heave | – | 0.1 g. |

The aluminium alloy selected for the tanks was 5356-0. Careful quality control of welding procedures was called for because

> 'Aluminium alloys are inherently somewhat difficult to weld well, particularly as regards porosity. Therefore it is essential that adequate procedures for quality control of the welding be set up to ensure that good welds are laid and that the completed welds are sound.'

100 per cent X-ray examination of all shell welds was called for with a hydrostatic test after completion, plus a 4 psi pneumatic test both after completion and after installation in the ship. In the event an overall weld reject rate of 26 per cent was achieved, which compared favourably with subsequent early independent tank designs but which has been vastly improved upon in recent constructions where less than 0.1% is now commonplace.

Roll and pitch locating keys were fitted at the top and bottom of each tank, the former fitted into stainless steel female keyways bolted to the underdeck girders, the latter into slots in the bottom insulation.

The specific gravity of the cargo was taken at 0.6 to enable LPG cargoes to be carried after the LNG trial period had been concluded.

**Cargo Handling:** The main cargo pumps comprised a single deep well pump of 1,000 gpm capacity in the centre of each tank; these discharged to a common deck header from which two booster pumps assisted in transferring cargo ashore.

The standby pumping equipment consisted of a 'blow-case' pump (fig. 6c) fitted in each tank – the philosophy being to install a unit of 'utter reliability and minimum working parts below deck' variety. This essentially comprised a small cylindrical vessel on the tank bottom – with non-return valves allowing LNG to flow into and fill the container; subsequently LNG vapour was

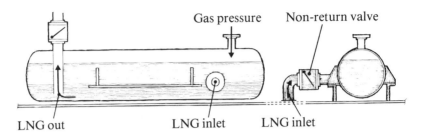

Fig. 6c. Blow case pump.

introduced into the vessel, forcing the LNG out. Simple, slow, inefficient – but it worked.

Level gauges were of the float type, the float being guided by three stainless steel wires and, in addition, being protected against cargo sloshing by a 'still well' tube. These were augmented by a 'Lucite' gauge fitted in the top of each tank which could be viewed through a specially designed sight glass (see fig. 8f).

All deck piping was of stainless steel (TP 304 L) with bellows expansion joints. Insulation was 'Foamglas' with mastic joint seals and coating.

The vapour relief system was rather conventional using weight loaded pressure/vacuum relief valves, for reliability, the pressure relief valves leading into a common header and riser.

Inert gas for the spaces around the tanks was supplied from a 2,000 gallon vacuum insulated liquid nitrogen storage tank fitted in the forward hold.

**Instrumentation and Controls:** The temperature of the inner hull plating and top keys was monitored by 68 thermocouples distributed over the surfaces, with audible alarms to sound at 20° F.

Gas sampling points were fitted in each hold space capable of detecting the presence of hydrocarbons in proportions of 0-50,000 parts per million.

No attempt was made to burn the cargo 'boil-off' in the main machinery which was diesel and unsuited to conversion; it was simply to be vented to atmosphere through the 30 foot high riser.

The conversion was duly carried out during the first half of 1958 at the Alabama Drydock & Shipbuilding Co., Mobile, Alabama – the tanks being built 100 miles up-river by the J. Ray McDermot Co. at Morgan City, Louisiana and shipped down by barge. Supervision was carried out by Lloyd's, ABS, British Board of Trade (represented by Lloyd's), USCG, Constock, British Gas (part time), J. J. Henry and many others – and was probably one of the most closely scrutinized merchant ships built this century!

## THE TRIALS

Trials of the ship, now renamed *Methane Pioneer*, commenced in October 1958:[2] these were planned to cover the initial inerting and filling of the tanks with liquid methane at the Lake Charles, Louisiana, site and to continue with sea trials in the Gulf of Mexico, concluding with inspection and evaluation of performance at Lake Charles. The time allocated was 83 days.

To quote from the Test Agenda:

'the tests have two purposes
(A) To determine the performance of tanks, insulation, cargo handling equipment and ship structure with liquid methane cargo
(B) to educate the ship's crew and others involved with respect to special features and operating procedures, so that comprehensive operating instructions may be prepared.'

The company which had been selected to operate the *Methane Pioneer* throughout the trials and subsequent transatlantic trial shipments, assuming all went well during the trials, was Stephenson Clarke & Co. Ltd. of London –hitherto more accustomed to shipping coal and oil fuel around the British Isles than liquefied natural gas over the high seas.

There is very little to report on the trials at Lake Charles which progressed smoothly and according to programme with few incidents of any real interest.

After progressively filling each tank, checking instrumentation and pumps (which did tend to stick due to differential expansion at the top seal/outlet fitting) and a jettison test over the stern, the ship was duly warmed up for inspection.

'Conch Photos'

51

Plate 1. *Methane Pioneer*–cargo jettison tests.

Natural Gas by Sea

One of the principal areas of concern was the adequacy of the cargo tanks and the aluminium welding, and it was this area which was subjected to the most critical examination after the trials. Difficulties had been met in agreeing acceptable standards of porosity, integrity of welds subjected to repeated cut-out/reweld operations, and particularly in the assessment of the hazard or otherwise presented by the small crater cracks which are difficult to eliminate with aluminium welding.

The inspection was carried out by a team of ten senior engineers – two per tank – although a total of no fewer than sixteen were directly involved at one stage. A total of 100 man days was expended on examination and carefully recording defects '. . . making use of magnifying glasses and flashlights' on over a total of 6,500 lineal feet of welding.

Among these Sherlock Holmes's were:

> on tank 2 – Dr. Fred Radd, (subsequently Chief Metallurgist for Conoco)
> on tank 3 – Professor 'Cheddy' Sliepcevich, (Oklahoma University)
> on tank 4 – D. Rooke, (subsequently Sir Denis Rooke, Chairman of British Gas Corporation – now retired)
> on tank 5 – Tony Rogan, (for the British Board of Trade)
> – Al Pastuhov, (subsequently President of American Technigaz)
> on tank 6 – Barry Thomas, (of the J.J. Henry Company; now compiler of LNG Log* for SIGTTO)

The weld crater cracks, of which there were some thirty or forty in each tank, formed the subject of considerable attention and discussion as to the likelihood of their propagating in service. The fact that they were confined to the fillet welds reduced the risk, but one member of the evaluation team remained unconvinced – to the extent that he pronounced that the ship would never reach the other side of the Atlantic!

After some persuasion, however, and reference to recent studies carried out by the British Welding Research Association on similar alloys, his fears were allayed. The most serious cracks were chipped out using '. . . cold chisels and gouges, honed to razor sharpness' and rewelded.

Thus, after '...an inspection that was nothing short of fantastic', the group unanimously agreed that the overall situation was satisfactory and the ship was pronounced fit to continue her planned programme of trial shipments of seven transatlantic crossings.

* Ref. 8 Ch.16.

Plate 2. Balsa lining completed.

Plate 3. Two cargo tanks arrive at Mobile.

Plate 4. Tank being lowered.

All pictures Conch photos.

Plate 5. One tank fitted in place.

Plate 7. *Methane Pioneer* after conversion.

Plate 6. Cargo tanks in transit.     Plates 2–7. *Methane Pioneer* conversion.

53

## THE FIRST TRANSATLANTIC CROSSING

*Methane Pioneer* left Lake Charles with a full cargo of LNG on her first historic 5,064 mile voyage on 25th January 1959; with an average speed of 9.4 knots, some heavy weather in mid-Atlantic and fog in the English Channel, she did not appear out of the mist at Canvey Island until the morning of 20th February to an excited and perhaps rather relieved reception party.

During that first trip, at one time rolling up to 30° (a 45° roll was recorded on a later crossing), she demonstrated to the complete satisfaction of all that the transportation of LNG by sea was a practical proposition—at least from a technical point of view; the subsequent six trips served only to consolidate this opinion.

The ship was fully warmed up on her return to Lake Charles, her tanks inspected and found in good order.

During this and subsequent trips considerable data was amassed on such items as daily boil off, temperature gradients, cargo behaviour (no strange phenomena in relation to the manner and mechanics of boiling were observed —or heard), cooldown and warm up procedures and times—all essential preliminaries to the design of a commercial scale venture.

It was during this first inspection period that Shell Petroleum arranged for a 'look see' at the Constock technology—a stake in Constock now would (a) enable them to recover two years' 'lost ground'; (b) satisfy British Gas's wish to have a major oil company involved; and (c) relieve the burden of Constock's future development costs. They must have liked what they saw because a few months later an agreement was signed under which Conch International Methane Limited was formed, with their day to day operations concentrated on Conch Methane Services Limited, London. This new company then comprised Continental Oil (40 per cent), Union Stockyards of Chicago (20 per cent), and Shell Petroleum (40 per cent).

For the more curious, the name Conch is an acronym composed from *Con*tinental and *Chi*cago, the whole word, Conch, being a univalve *Shell*. It was conceived by the fertile mind of one of the members of the then Shell Natural Gas Division. Thus the theory of a 'friend' of the author that the *Con* derived from *con*fidence trick is hopefully hereby, and forever, disproved.

The completely successful transatlantic crossing of the *Methane Pioneer*,* and the cooldown and proving of the Canvey Methane Terminal in 1959, was

---

* *Methane Pioneer* was sold to Gazocean in 1965 after a period of LPG service and renamed *Aristotle,* remaining in the LPG trade until 1973 during which period she also delivered fifteen LNG cargoes to Barcelona (1969) and six to Boston (1968-71). She was subsequently sold to Coperbo of Recife, where she was last reported in 1988 as still being used for butadiene storage.

the date of the birth of Natural Gas by Sea. No real obstacle now remained to prevent the rapid and unfettered expansion of this new and exciting technology. Let there be no doubt that the excitement, even amongst the most sober-minded of men, was intense. There seemed no limit to the potential expansion of the LNG market for, after all, were not the world gas reserves far in excess of all the oil discovered so far? A clean fuel for an age of ever increasing pollution and a proven technology for shipping it across the high seas presented a glittering prospect indeed.

[1] 'Ocean transport of liquid methane' by John A. Murphy and C. G. Filstead. 5th World Petroleum Conference, 1959.
[2] 'Low-temperature, liquefied-gas transportation' by C. G. Filstead and Montgomery Banister. SNAME, Vol 69, 1961.

# 7
# The Second Prototype –
# *Beauvais*

LNG ship development work in France took an interestingly different form to that in the USA.

Whereas the American work was concentrated in a single group, working along a single, and as far as possible pre-planned, path, in France the work, although stimulated by Gaz de France through the Worms Group, was to a great extent shared between a number of companies.

With M. Labbé's return from the USA in June 1959, determined to 'make a study', and the formation of the Methane Transport Company to co-ordinate all work in France, real progress began to be made. Methane Transport comprised Gaz de France, Worms, Air Liquide, Segans, several banks and also the Gazocean shipping company; it worked closely with the four principal shipyards, Chantiers de l'Atlantique, Ateliers et Chantiers de la Seine Maritime (Le Trait), Les Ateliers et Chantiers de Dunkerque–Bordeaux, and Forges et Chantiers de la Méditerranée (La Seyne)–now CNIM; the shipyards were not, however, partners in Methane Transport Company.

Each shipyard had, since about 1957, been making its own study of LNG, stimulated by the worldwide interest in this commodity and the enouragement of Gaz de France, the more so as it was the shipyards' stated policy at this time to concentrate on the more sophisticated type of ship in preference to the mass production of conventional oil tankers.

# 'BEAUVAIS'

## (i) GENERAL VIEW

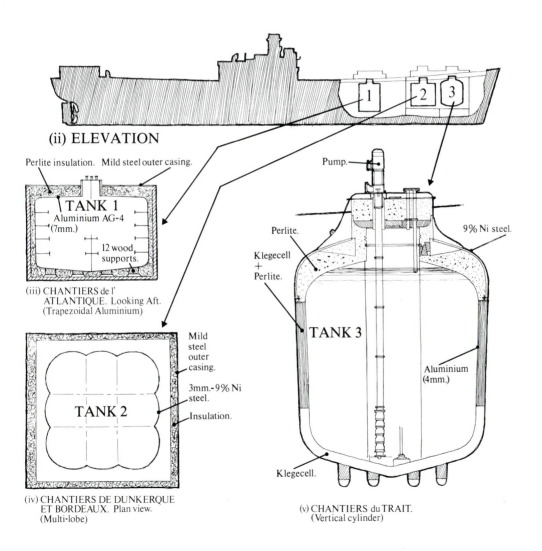

## (ii) ELEVATION

Perlite insulation.   Mild steel outer casing.

TANK 1

Aluminium AG-4
(7mm.)

12 wood
supports.

(iii) CHANTIERS de l'
ATLANTIQUE. Looking Aft.
(Trapezoidal Aluminium)

Mild
steel
outer
casing.

3mm.-9% Ni
steel.

Insulation.

TANK 2

(iv) CHANTIERS DE DUNKERQUE
ET BORDEAUX.  Plan view.
(Multi-lobe)

Pump.

Perlite.

Klegecell
+
Perlite.

9% Ni steel.

TANK 3

Aluminium
(4mm.)

Klegecell.

(v) CHANTIERS du TRAIT.
(Vertical cylinder)

Thus the development of LNG ship technology progressed on an industry-wide basis; and in order to reach a preferred design at the earliest date it was concluded that a prototype ship, incorporating a variety of containment systems, should be fully tested and evaluated under seagoing conditions.

To facilitate this endeavour the French Government made a wartime-built Liberty ship, *Beauvais,* available at a nominal price.

## THE CONVERSION[2]

The *Beauvais* conversion was arranged, and responsibilities allocated, in the following manner:

**Containment Systems:** (see fig. 7a)

Tank No. 1: Atlantique design. An aluminium tank, 400 m³ capacity, of self-supporting, prismatic* form. The internal structure was conventional ship-type; material, aluminium alloy AG-4.

Insulation comprised 850mm PVC foam blocks, attached to the mild steel outer shell (simulating the ship's inner hull), faced on the 'cold' side with glass reinforced polyester resin to act as a secondary barrier (three different 'layups' were used). Perlite was used to fill the space between tank and secondary barrier.

This tank was dimensionally half-scale as compared with a projected 25,000m³ commercial ship tank, giving a volume ratio of approximately 1:8.

Tank No. 2: Dunkerque and Bordeaux design. Of multi-lobe or poly-cylindrical form; 120m³ capacity constructed of 9 per cent Ni steel and located against rolling at its central axis. The tank was enclosed in a mild steel outer container and insulated, it is believed, by PVC foam block insulation.

Tank No. 3: Chantiers du Trait. Of vertical cylindrical form, supported at its axis; 120m³ capacity and constructed of AG-4 aluminium alloy. The outer container was 9 per cent Ni steel; the insulation, part expanded PVC blocks, part Perlite. Chantiers de la Méditerranée also contributed to this tank design.

**Cargo handling:** specific design and operating responsibilities were divided between the yards as follows:

　　　Le Trait　　– methane vapour system

---

* The French use the almost unpronounceable word 'parallélépipédique' to describe this type of tank!

57

Plate 1. Tank 1. Applying the polyester resin/glass cloth secondary barrier to the face of the PVC block insulation; each wall of the outer casing was prefabricated – the corner insulation being completed in situ on board.

Plate 2. Tank 1. Internal view – showing 'heating' coils for vaporizing 'unpumpables' also lower end of Nitrogen bubbler level gauge.

Plate 3. Tank 1. Internal structure – showing instrumentation.

Plate 4. Tank 3 being lifted into position.

Plate 5. Tank 3. Showing bottom locating keys and tank supports – note also sintered metal 'pads' for fluidizing Perlite insulation.

Plate 6. '*Beauvais* after conversion' – note deckhouses over experimental tanks and vent risers attached to masts, terminating well above deckhouse level.

Dunkerque – controls and measurement

La Seyne  – methane liquid handling

Atlantique – overall co-ordination of conversion and sea trials.

As to the equipment itself, there were a number of interesting items incorporated into the *Beauvais* installation which constituted 'firsts' as far as marine application was concerned.

**Pumps:** Different systems were installed:

(a) deepwell (Guinard);

(b) fully submerged electrical (J. C. Carter);

(c) an eductor combined with deck mounted pump (Segans);

(d) a gaseous piston pump designed by M. Guilhem.

**Level Gauges** were of

(a) capacitance type;

(b) bubbler type using nitrogen;

(c) radioactive type.

The float type adopted by the USA and available at that time (but not in France) does not seem to have been fitted.

**Gas Detection/Temperature Measurement:** Conventional equipment, e.g. thermocouples and infrared gas detection, was available and used. However, in the matter of gas detection Gaz de France clearly had no faith in existing instrumentation and insisted that at least ten canaries '. . . ready to pay with their lives in the event of the smallest leakage of methane . . .' be carried on board throughout the tests.

The fact that seafarers are, by nature, sentimental towards animals, and were inclined to transfer the birds from the cargo holds to their quarters, was a relatively minor complication in the overall programme.

**Piping and Valves:** Expansion bends were used wherever possible, minimizing the number of bellows.

The conversion was commenced in March 1961 and completed in February 1962, under the supervision, and to the full approval of Bureau Veritas, although Lloyd's Register of Shipping was also involved in all stages of the work in an advisory role.

## THE SEA TRIALS

The cooldown, followed by sea trials, commenced in March 1962 and extended over a period of about six months. The programme was as follows:

Phase 1: alongside at St. Nazaire; cooldown tests with liquid nitrogen.

Phase 2: alongside at Roche Maurice; loading and discharging (LNG) into and out of each tank and checking the associated installations.

Phase 3: at sea; series of trips with tanks 2 and 3 full and 1 empty as in 'ballast' condition.

Phase 4: at sea; series of trips with tank 1 full and tanks 2 and 3 empty as in 'ballast' condition.

Phase 5: at sea; alternating transfer of cargo between tanks (after warming up empty tanks); these tests were to check their capability of withstanding repeated thermal shocks.

Phase 6: at sea; tests of gasfreeing and warming up tanks.

Between each phase *Beauvais* was to return to Roche Maurice to adjust cargo for the following phase.

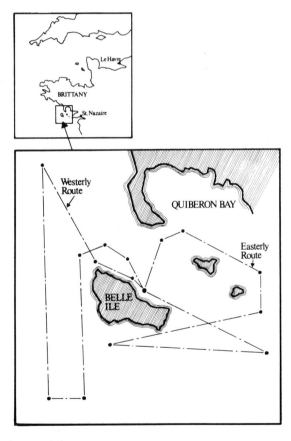

Fig. 7b. *Beauvais* sea trials.

In all, the trials extended over a period of four months. One of the main problems was the lack of sufficiently bad weather over the entire period to obtain useful measurements of the effect of ship movements, and sea water on the deck equipment–although the latter could be, and was, simulated with hoses.

Boredom in the experimental team was relieved by allowing each member to take his turn as navigator and helmsman for a day. As to the canaries, nature seemed to take over for, in the words of M. Chemereau of Atlantique, himself a team member, '. . . these birds felt very well on board and their population was at least trebled at the end of the tests!' which, apart from any other interpretation one may wish to make, at least attested to the fact that there must have been a conspicuous absence of methane in the atmosphere.

One of the few surprising incidents was the failure of the submerged pump after a few revolutions. During the inspection after warm up, it was found that the stainless steel separator in one of the bearing ball races had deformed and caused the ball race to lock, resulting in the disintegration of the bearing. In fact an unauthorized substitution of stainless steel for the approved epoxy laminate had been made by the bearing manufacturers. Reverting to the original approved separator material resulted in completely satisfactory results thereafter—further confirmed by the fact that it is still used in the much higher capacity cargo pumps of today.

As to the overall technical appraisal of the trials, generally speaking '. . . almost to the surprise of the technicians . . .' the tests produced no incident of any importance; and as M. Gilles wrote in an article published in *Nouveautés Techniques Maritime* in 1963:

> 'they confirmed the existing opinion of many people, namely that there was a number of ways to design a satisfactory LNG containment system. The operation proved, above all, one basic fact, that no design, however ingenious, has any value unless it has been built and tested. Furthermore, it can be concluded that the construction of a methane ship required considerably more technological "knowhow" than patented ideas. One of the essential results of the *Beauvais* operation was to enable the shipbuilders concerned to acquire this technological "knowhow".'

This was, in fact, an element which was completely missing from the *Methane Pioneer* exercise and must have contributed in large measure to the dominating position which the French yards were able to take in later years.

[1] 'Le méthanier expérimental *Beauvais*' by A. Gilles. Extract from *Nouveautés Techniques Maritimes*, 1963.
[2] Gaz de France information booklet No 162, 1st September 1962.

# 8
# Cargo Handling Arrangements

This chapter is concerned with the problems presented and solutions adopted in loading and discharging the cargo and, perhaps even more important, maintaining and monitoring the design conditions on board ship both at sea and in port. In the longer term it would be necessary to devise a safe and economic method of using the cargo 'boil off' which, on both prototype ships, was vented to atmosphere and wasted; its potential as a very desirable fuel for the ship's main machinery was undisputed.

Because LNG was to be carried as a boiling liquid it could be expected to present handling problems – due to vaporization in the pumps, vapour locks in the piping, and differential expansion of many items of equipment. What materials should be used, for example, for the joints at flanged connections? How would the bolted flanged joints themselves react to temperature changes? How 'searching' would LNG prove to be as compared with, for example, gasoline?

In selecting and, in most cases, developing the necessary equipment the essential requirements were *safety* and *reliability*.

With Cleveland still quite fresh in everyone's mind it was of paramount importance that every part of the system, however small and seemingly insignificant, should be carefully evaluated and engineered, then rechecked and checked again.

**Procedures:** It had been decided from the outset that the cargo would be maintained under a small positive pressure at all times – thus ensuring that no air entered the containment, or piping, systems at any time, and this included the loading and discharge operations.

To maintain this positive pressure at sea presented no problem because the continuously boiling cargo provided this precise condition automatically – subject, of course, to proper controls to prevent over-pressure.

During loading and discharge operations, however, excess vapour would have to be returned to shore or delivered into the ship's tanks respectively and in carefully controlled quantities.

During the initial cooling down of the tanks, itself a delicate operation in order to avoid stress concentrations due to unequal differential contraction of the tank surfaces, the large amount of vaporized coolant, had to be disposed of.

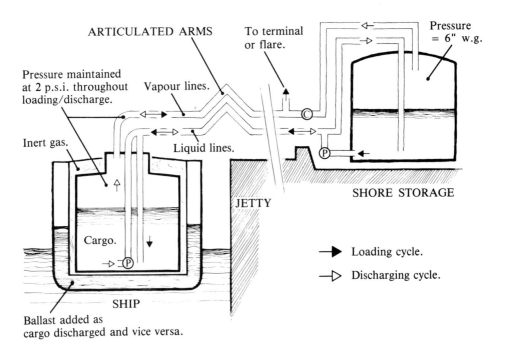

Fig. 8a.  Closed cycle Loading and Discharge.

As an extra precautionary measure it was established that, as a basic policy, at no stage in the ship's operating life would flammable gas/air mixtures be allowed to exist in the ship's tanks – or, indeed, in the space around them (unlike normal practice in conventional oil tankers at that time). This involved purging air from the tanks with an inert gas before loading and, similarly, purging the gas from the tanks before aerating them for annual inspection. Furthermore, vacuum relief would be provided by supplying gaseous nitrogen to the tanks rather than air.

The spaces around the tanks – or the 'insulation' or 'hold' spaces – were also to be maintained in a 100 per cent inert condition; quite apart from the safety aspect this had the additional advantage of keeping the insulation moisture free. In reality it was this latter consideration which was the prime reason for the inerted hold spaces because it would have been quite simple to ensure that these spaces, if air filled, were maintained below the flammable limit by conventional means, viz. gas detectors combined with adequate forced ventilation. As operating experience has been gained over the years it has been possible to relax the requirement for an inert gas atmosphere around the independent type 'B' tanks; these spaces are now maintained under a small positive pressure of dry air – a less expensive solution – which also permits easy access for inspection of these spaces in service.

The basic 'cargo handling' design parameters were thus established as follows:

(1) Cooldown and warm up procedures designed to avoid unacceptable temperature gradients.
(2) Cargo discharge equipment of great reliability. No bottom, or indeed below deck, connections to the tanks.
(3) Means to monitor temperature levels in the tank structure and the hull structure of the ship.
(4) Means to measure liquid levels in the tanks.
(5) Means to monitor for tank leakage.
(6) Means to control and measure vapour pressure in both tanks and deck piping systems under all service conditions.
(7) Provide for expansion and contraction of piping caused by thermal changes and ship structural movements.
(8) Means to prevent or counteract cooling of hull structure in the event of local insulation failure.
(9) Means safely to dispose of or jettison cargo from the tanks themselves, or the spaces around them, in the event of a major tank/insulation failure.
(10) Means safely to collect and burn in the ship's main machinery installation the cargo boil off

and, but later,

65

(11) Means to detect and dispose of water leakage from adjacent ballast spaces into the LNG containment spaces.

and later again,

(12) Means to prevent venting methane vapour to atmosphere under any foreseeable circumstances whatsoever. (The requirement of over-sensitive Port Authorities.) This unnecessarily cautious requirement was subsequently relaxed to permit methane vapour to be vented to atmosphere in an emergency, provided that it was sufficiently warmed to ensure that its density was less than that of air.

The selection and development of materials and equipment to meet these requirements resulted in some interesting ideas in the early days. However, in direct contrast to the containment systems themselves which still vary very considerably in both concept and design, the cargo handling procedures and equipment for LNG ships acquired a reasonably standard format quite quickly.

**Pumps:** It will have been noted in the preceding chapters describing the *Methane Pioneer* and *Beauvais* that there was some variation in the installed pumping equipment. By and large, existing pump types were used, and these were, it must be said, a qualified success.

Because of the time and expense involved in warm-up, inerting and subsequent cooling required to gain access to any one tank, the reliability of these units was clearly vital to the economic operation of a commercial vessel and, as a result, all early proposals included an 'utterly reliable' standby system which had, effectively, no working parts below deck. All such systems, though effective, were clumsy and inefficient.

Largely as a result of the problems encountered with differential expansion and binding of the drive shaft of the deepwell pumps on *Methane Pioneer* which, it was felt, could only become worse as ships grew in size, attention was directed towards the use of a fully submerged, electrically driven, pump. This was seen as a 'safe' concept for LNG because the cargo tanks were, as already described, maintained at a positive pressure at all times, thus excluding air and, with it, any possible existence of a flammable mixture in the tanks. The only apparent alternative–bottom tank connections and pumprooms below deck, as used on conventional tankers–was considered to be too unattractive and complicated to be worth serious consideration.

Tests on a 2hp submerged pump, manufactured by the J. C. Carter Co. of Costa Mesa, California, were already being carried out in 1960 at Lake Charles, La., as part of Constock's ongoing research into the inground storage of LNG. The pump, a standard unit, had performed well on test, and after clocking up some 700 running hours in LNG, sufficient confidence was gained to make overtures to the ship classification societies for approval of their use in LNG ships. After the initial shock of contemplating the presence of electric cabling and motors in a tank full of hydrocarbon liquid, the logic of the proposal, backed by hard facts on, for example, the electrical resistance of methane and the ability to ensure the absence of oxygen from the cargo tanks at any time, overcame initial reluctance; first Bureau Veritas (for experimental purposes only) then Lloyd's gave their approval–subject to certain specific operational precautionary measures. ABS followed a few months later (1962) after a somewhat protracted rearguard action by one of the older and more conservative members of their Technical Committee.

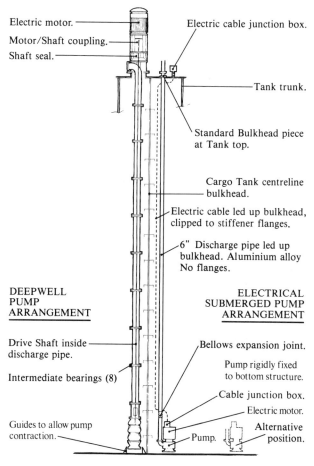

Fig. 8b. Alternative Pump arrangements for U.K. ships.

The first submerged pump unit was duly installed in November/December 1961 on *Beauvais*–and duly failed after a few revolutions! As already mentioned, this was found to be due to the late and unauthorized substitution of the standard epoxy laminate separator in the ball bearings by an untested stainless steel plate which distorted and in turn destroyed the bearing. A new set of bearings solved the problem and no further trouble was experienced.

Similar pumps were subsequently fitted to *Methane Princess* and *Methane Progress* with complete success; and a year later, to *Jules Verne*.[1] Present day separators are either epoxy laminate or specially modified nylon.

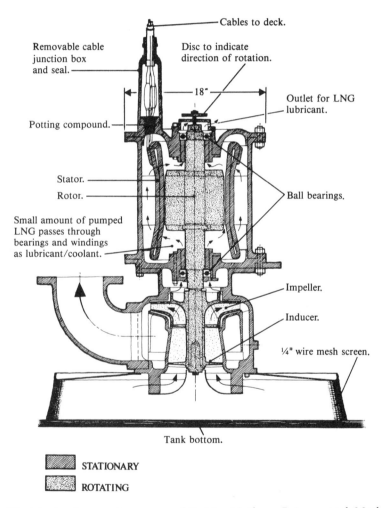

Fig. 8c. Electrical submerged pumps as fitted to *Methane Princess* and *Methane Progress*

Plate 1. 2 H.P. Electrical submerged pump tested at Lake Charles (1960–1)

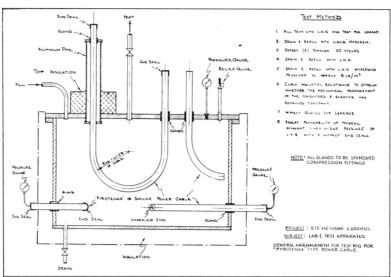

Plate 2. Test proposal for cable end seal techniques.

Plate 5. General view of '*Princess*'/'*Progress*' pump.

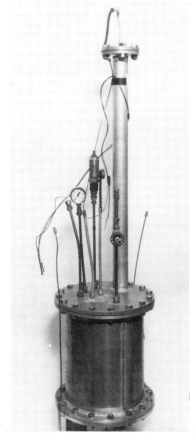

Plate 3. Test rig for cable end seal.

Plate 4. Production tests of pumps for '*Methane Princess*'/'*Methane Progress*'

One of the attractive features of these pumps is the 'inducer' which is the key to their very low NPSH, or pumpdown, characteristic; in fact, the *Methane Princess/Methane Progress* pumps were able to draw the cargo down to about 6 inches even with the disadvantage of the bottom stiffening structure interfering with the free flow of cargo into the suction.

In addition to the careful prototype testing of the pumps themselves it was very important to be certain that the cabling to and from the pumps was secure. Special attention was paid to the top and bottom end seals; it was feared that if liquid methane should migrate into the mineral insulation core while the tanks were full and cold, its rapid expansion during warm up prior to drydocking could cause rupture of the copper sheathing. Furthermore, leakage through the gland at deck level could cause either methane leakage to atmosphere or moisture ingress into the cable–both equally undesirable.

A comprehensive test programme was set up at Shell's Thornton Laboratory to select suitable jointing techniques. Similar work was carried out independently in France.

Several ideas were evaluated on the early prototype and commercial tankers as a back-up to the then 'experimental' electrical submerged pump.

The blow case pump design on *Methane Pioneer* has already been described in Chapter 6.

On *Beauvais,* apart from testing different types of pumps, an eductor was fitted as a means of discharging 'without any working parts below deck'.

On *Methane Princess/Methane Progress*, a 'vapour lift' system (Shell patent) was fitted to each tank–and the adjacent hold spaces; this employed a technique requiring a vacuum/separator vessel on deck connected to each tank with a single pipe. By reducing the pressure in the deck vessel a vapour/

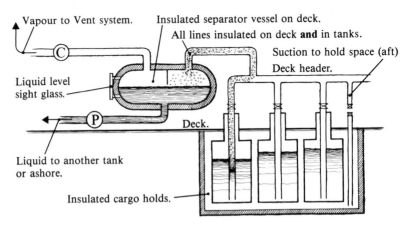

Fig. 8d.  Principle of Vapour Lift System.

liquid mixture was drawn up into the separator from which the liquid was pumped into an adjacent tank, or ashore, by conventional deck mounted pump(s). The advantage of this system was that it could also be connected to the very narrow (3 inch) space around the tanks to remove LNG leakage into the secondary barrier space, for which purpose incidentally, it has not yet been used!

Again, a simple and inefficient system, but it worked.

Today the problem is solved by the simple expedient of fitting two main cargo pumps in each tank plus a small emergency/stripping pump – all of the electrical submerged type. Their reliability is now well proven.

**Ship/Shore Transfer:** Tests on hoses had failed to find a suitable design for LNG service; attention was therefore directed towards the articulated pipe design – already well established for conventional oil tanker operations.

It was not long before a suitable swivel joint using Teflon seals was developed and successfully tested by both *Beauvais* and *Methane Pioneer*. Leakage at the joints was minimal and the technique has since been used for all LNG operations with entirely satisfactory results. The arms can be either aluminium or stainless steel, and are generally left uninsulated.

**Deck Piping/Valves, etc.:** The use of stainless steel was imposed by Classification Societies because of the additional fire risk with aluminium which has a much lower melting point; in any case stainless steel is easier to weld; it has other advantages, not least its lower coefficient of thermal expansion.

There already existed, in the USA, a substantial body of experience with stainless steel cryogenic piping systems. It only remained to adapt this to shipboard use – taking special account of the marine atmosphere (selection of alloys) and the continuous movement of the ship structure to which the system would be attached.

Particular emphasis was placed on maximum simplicity and reliability, for example, initially in the use of weight-loaded relief valves (no springs or diaphragms to fail) although subsequently more conventional spring loaded valves were introduced. Much careful design study was also put into flanged connections, for it was extremely important that these did not leak (fig. 8e). In the event it was found that standard asbestos joints were entirely adequate and that, provided the bolts were tightened by torque wrench to a defined pre-tension, this compensated for differential shrinkage, even at the tank top fittings where aluminium flanges were coupled to stainless steel.

Attention was also given, of course, to the analysis of the piping system as a whole, particularly in regard to defining the total movements to which each section would be subjected over its working life–taking into proper account both thermal and ship structural movements.

Valving was confined to types with known performance at low temperature, for example gate valves; however, ball valves and–much later–butterfly

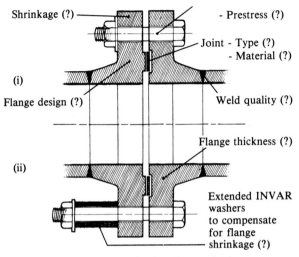

Fig. 8e. Some problems associated with pipe flange design.

valves, were subsequently found to be satisfactory. In the case of valves it was essential that (a) no liquid could be trapped in enclosed spaces–causing rupture on expansion; this also applied to 'closed' sections of piping which were provided with separate relief valves; and (b) that they should be 'fire safe'–which meant that, in the event of fire destroying the Teflon seals, the valve 'gate' would be supported by a metal back up ring on the 'down stream' side.

**Insulation:** The main problem with deck piping insulation is to ensure that water is kept out of the system and that access is available for tightening up flange bolts; achieving effective water tightness at the pipe supports is particularly difficult.

Inorganic, closed pore, preformed foam blocks were selected in the first instance and in fact proved to be entirely satisfactory in service, though expensive; preformed polyurethane foams have been used with success since. In all cases a bitumen based surface mastic coating, reinforced with glass cloth or similar was used on the prototype ships–and this practice has continued to the present day.

**Expansion/Contraction Arrangements:** It was considered that steel bellows were the only acceptable solution. In fact, there was already much experience with bellows in the USA in 1958/9. It merely remained to define the physical movements which would be experienced–and their frequency; then to establish a qualification test programme and manufacturing procedure. No problems have been experienced with these items provided they are properly fitted.

The need for bellows expansion pieces, which are not universally liked, can often be much reduced by arranging for the piping to deflect through long-legged angled bends.

In total, it can be said that the careful approach to the design of the cargo handling equipment has resulted in a rather simple and reliable solution which has remained relatively unchanged throughout the short history of LNG transportation.

**Instrumentation and Control Systems:** This was, from the start, seen to be one of the principal keys to safe and reliable operations, and reliability of the equipment itself was of vital importance because much of it would be inaccessible for maintenance for the period between survey and drydockings; some items, such as temperature sensors, would in fact be 'buried' for the life of the ship.

**Temperature Measurement:** Equipment was needed to provide essential information in three areas, all directly related to the safety of the ship:

— inner hull temperatures – so as to monitor the effective performance of the insulation and secondary barrier protection system; particularly in areas such as keyways and tank supports.
— tank temperatures–to monitor the temperature gradients in the tanks during the cooldown and warm up operations
— to detect the presence of liquid, as opposed to gas, in the spaces around the tanks in the event of tank leakage.

Information on the temperature of the cargo itself was also required as an essential ingredient for the formula which derives the calorific value of the cargo, or for what is now termed 'custody transfer'. Many of these temperature measurement devices were required to detect quite small temperature changes and over a wide range, and to transmit this information over a considerable distance and to activate alarms in certain cases.

The question of the required spacing of inner hull temperature sensors to achieve an adequate indication of local insulation breakdown was debated for a long time—until finally, in 1963, E.Abrahamsen, of Norske Veritas[2] persuaded us all that there was no real point in trying—better to spend the money on better quality steel!

**Liquid Level Measurement:** This was clearly an important matter, not only for custody transfer purposes, but to establish when the tanks were completely filled – or empty.

All design groups were agreed that the reliability of the system was of paramount importance for safety reasons if no other; overflowing a tank of gasoline or crude oil was one thing, but a tank of LNG was a very much more serious matter. It was judged to be prudent therefore to have a reliable back up system, or even two.

In fact *Methane Pioneer* had two and *Methane Princess* and *Methane Progress* three back up systems; the *Jules Verne* had two (which led to trouble as will be seen in a later chapter).

Float gauges were the obvious first choice – these were already in use on

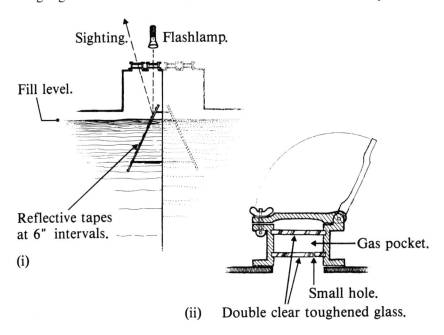

Fig. 8f. Arrangement and details of sight glasses and gauge boards.

pressurized LPG ships and simply required modifying in respect of materials, for cryogenic applications. Nonetheless some initial problems were experienced with sticking floats and broken tapes–the latter putting the whole equipment out of action until the next tank warm up as there was no way of retrieving the float.

Back up systems on the *Methane Pioneer* were simply sight glasses in the trunk top together with a 'Lucite' reflective gauge board which extended over the top few feet of the tank; as a secondary measure, there were temperature indications on the tank wall, but this was relatively inaccurate. The sight glass and gauge board was found to be an extremely effective and totally reliable back up device, and was subsequently fitted to *Methane Princess* and *Methane Progress* with complete success.

As a further protection against over-filling the tanks, alarms and auto shut-off devices were fitted to the float gauges on *Methane Pioneer* – but such was the importance attached to the need to prevent overfilling that subsequent ships, including *Methane Princess* and *Methane Progress*, incorporated devices which were totally independent of all other equipment; this has now become mandatory and, indeed, makes good sense. The first of such pieces of equipment were based on the principle of electrical capacitance – indeed such a fitting was incorporated on *Beauvais*: their initial weakness was over-sensitivity, in that the early versions tended to react to the dense vapours which occur during the final filling stages, thus producing frequent false alarms and auto shutdowns of the loading operation, further resulting, on one occasion, in its disconnection by a frustrated cargo officer! The fittings have now been 'desensitized' and provide a reliable back up service.

Another system, fitted to *Beauvais* for level measurement and to later ships – and also used for density measurement, is the nitrogen bubbler system, based on the well tried pneumercator method of oil fuel bunker and ballast water measurement; this simply displaces cargo from a vertical tube led to the bottom of the tank and measures the resulting head of liquid outside the tube. It was found on trials, however, that it was impossible to obtain stable measurements with such equipment; the problem was eventually traced to the fact that nitrogen was quite soluble in LNG, the simple solution being to provide a constant *flow* of nitrogen and to measure the liquid head against this flow. This method is still used. It is not as accurate as more modern techniques such as ultrasonics, but does provide an alternative and relatively simple back up to the more sophisticated instruments.[3/4]

To use the nitrogen bubbler system for measurement of cargo density requires two tubes terminating a fixed distance apart–the differential head provides the density.

Natural Gas by Sea

To summarize, therefore, after a few false starts and what could be termed normal engineering refinements, a safe and reliable cargo handling arrangement, with its associated instruments and control systems, was developed within a few years of the first two prototypes entering service; there has been little change of substance since, which ever of the containment systems is used.

[1] 'Use of submerged electrical motor driven pumps for liquefied gases' by E. Hylton and R. G. Jackson. International Conference on LNG, organized by the International Institute of Refrigeration and the British Cryogenics Council, London, March 1969.
[2] 'Gas transport and ship classification' by E. Abrahamsen. *European Shipbuilding* No 2, 1963.
[3] 'The special cargo instrumentation requirements of LNG ships' by R. L. Blanchard, LNG 73, London, 1973.
[4] 'Measurement of density in custody transfer systems' by R. L. Blanchard. Gastech 75, Paris, 1975.

# 9

# The First Commercial
# LNG Ships

## *METHANE PRINCESS* AND *METHANE PROGRESS*

Events moved rapidly after the successful trial shipments of the *Methane Pioneer*.

Preliminary negotiations with Venezuelan gas producers had failed to achieve any prospects for long term shipments for the UK but by 1960 negotiations were well advanced towards the conclusion of a fifteen year contract with Algeria to ship 100 million cubic feet of gas/day to Great Britain and half that amount to France. Designs for the ships were being developed concurrently. The new Conch company was setting up its headquarters in London.

Although the British Gas Corporation was closely involved with the UK project and indeed was to take a 50 per cent share in the ship operating company, British Methane, and was also, through their North Thames Gas Board, responsible for the design and construction of the new reception terminal at Canvey Island, it was Conch who acted as the principals in this first project – providing both the initiating and central co-ordinating functions. Conch, through CAMEL (Compagnie Algérienne du Méthane Liquide), were also responsible for the design, and were to become 50 per cent owners, of the liquefaction and storage plant, and the marine terminal, at Arzew.

The *Methane Pioneer* trial shipment, together with Shell's input on ship operation and performance, had provided sufficient information to enable J. J. Henry to optimize the size and speed of the two ships required for the UK project.[1]

Adopting the following basic operational assumptions for a range of speeds:

Days in service per year .. .. .. .. 342
Port time per round voyage .. .. .. $3\frac{1}{2}$ days
Average boil off per day .. .. .. .. 0.33 per cent
Round voyage .. .. .. .. .. 3060 miles

together with such factors as

(a) shore storage costs,
(b) the effect of cargo boil off 'loss' and its value as fuel by way of compensation,
(c) port turn round and annual drydocking times which included due allowances for warm up, aeration, cooldown and inerting,

resulted in an optimized vessel with a loaded capacity of 167,000 bbls (26,450m$^3$), an average service speed of 17¼ knots, and power of 12,500 shp.

Specifications were then drawn up, and tenders sought – but only from UK

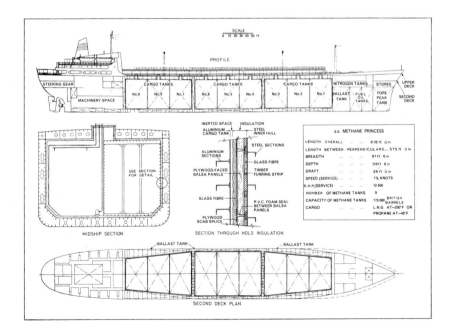

Fig. 9a. General arrangement of *'Methane Princess'* and *'Methane Progress'*.

shipyards (the second instance of politics entering the LNG scene).

At this time the Conch tank design was restricted to a rectangular shape; computer capacities could only just handle the introduction of a centreline bulkhead—now required for adequate transverse stability of the ship: this resulted in the rather ineffective use of the forward and aft tapered holds. However, within a few months a larger capacity computer became available and it was possible to consider tapered, or trapezoidal, shaped tanks; this immediately increased the capacity of the ship, whose dimensions were by this time (1961) fixed, to 170,000 bbls. Further studies on insulation thickness and tank/hold clearances resulted in a final capacity of 173,000 bbls (27,400 m³). The whole project had now moved forward to the selection of two shipyards (Vickers Armstrong and Harland & Wolff), preliminary model tests had been carried out and steel orders placed. The building contracts were signed in February 1962 along with the gas purchase, charters, subcharters and sales contracts, UK Government approval for the project having been obtained a few weeks previously.

Vickers Shipbuilders, Barrow-in-Furness, were allocated the responsibility of lead yard, thus originating all the working plans and material orders for both ships and obtaining classification and owners' approval; however, Harland & Wolff were free to adapt the conventional items to their own yard practices if they so wished, thus retaining some degree of flexibility.

In order to be certain

(1) that the experience obtained during the design and construction and operation of the *Methane Pioneer* should be passed on in the most efficient manner to the builders of the two new ships;

(2) that all the participants and contributors in the early work should be properly represented;

(3) that the eventual customers (British Gas) should be properly protected in regard to continuity of supply and cost of product;

an integrated procedure for plan approval and supervision was set up at an early date, from which it will be seen that Shell International's marine department played an active and important role in the construction of the two ships.

Every one of the 1,000 or more plans for each ship was approved by five separate organizations—two Classification Societies, Shell and/or J. J. Henry, Conch and British Gas; and, although this arrangement might seem cumbersome at first sight, the dedication and goodwill displayed by all parties ensured its sucess. The fact that

— the ships were built on time;

— they were built for within 1 per cent of the contract price;

— and they operated reliably for well over their original 15 year contract is perhaps evidence enough of the workability – perhaps even the desirability – of such an arrangement.

Final technical details of these two ships on which the building contracts were subsequently based were as follows:

| | |
|---|---|
| Length – overall | 621 ft 0 ins |
| Length – between perpendiculars | 575 ft 0 ins |
| Breadth – moulded | 81 ft 6 ins |
| Depth – moulded | 58 ft 6 ins |
| Draft – maximum design (scantling) | 35 ft 0 ins |
| Draft – normal LNG service | 27 ft 6 ins |
| Tank volume | 173,000 bbls. |
| SHP | 12,500 |
| Speed on trial (knots) | 18.5 |
| Service speed (knots) | 17.25 |
| Number of holds | 3 |
| Number of tanks | 9 |
| Material of tanks | 5083-0 aluminium alloy |
| Tank design stress (psi) | 12,000 – plating |
| | 10,000 – extrusions |
| Locating keys | top/bottom |
| Insulation | balsa/glassfibre |
| Boil off per day (average estd.) | 0.33 per cent of total volume |
| Relief valve setting | 3.0/3.5 psig |
| Overall discharge time | 18 hours |
| Overall loading time | 18 hours |
| Classification | dual, ABS and Lloyd's Register of Shipping |

The reason for the 'two' drafts was twofold – firstly, the ships were designed to be able to carry a full cargo of LPG – and so guard against the possibility of LNG cargoes being unavailable either for technical or commercial reasons; secondly, provision had to be made for the possibility of filling one or more of the ballast tanks with the ship fully loaded with LNG, and thus maintaining the inner hull structure at a safe temperature in the event of a major failure in the insulation system.

Dual classification was decided upon for two reasons; one, to acquire the maximum benefit from the combined experience of the two Classification Societies – both of whom had been involved closely throughout the development phase, and two, to ensure the maximum flowback of first hand experience

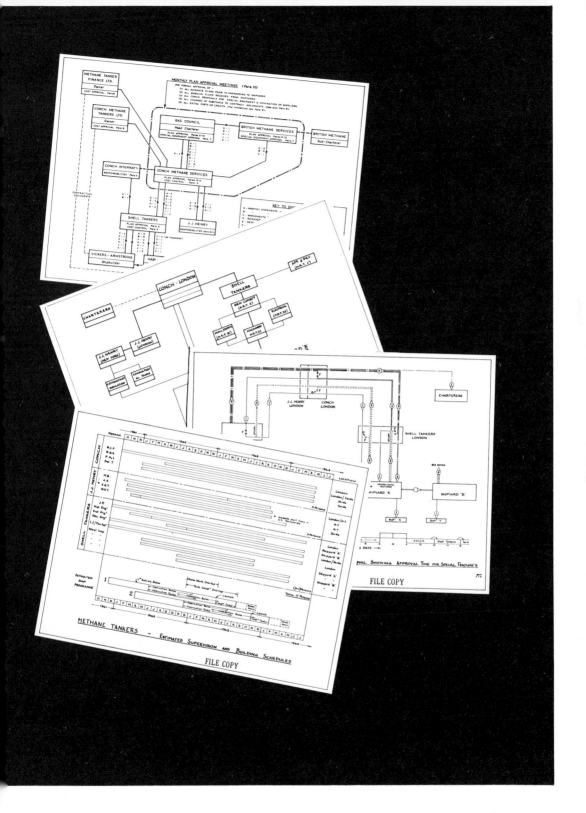

Fig. 9b.  Building and Construction Charts.

to industry – this latter might at first sight appear somewhat altruistic and even perhaps in conflict with Conch's patent policy to protect such knowhow, but, in the final analysis, their principal long term objective was for the LNG market to establish itself on a firm foundation from which it could grow and prosper.

A number of papers have been published describing in detail the *Methane Princess* and *Methane Progress* and their subsequent performance in service – these are referenced at the end of this chapter[2][3][4]; nevertheless there are a few matters relating to their design, construction and operation which are particularly relevant to the development of LNG ship technology and for this reason are considered to be worth recording here; they appear in no special sequence or order of significance.

**Aluminium Welding:**The welding of the *Methane Pioneer* tanks had presented many problems in the early stages. Careful quality control procedures and inspection were essential in order to obtain welds which

    (a) were acceptably free from porosity, and lack of penetration or inclusions;

    (b) had a reasonably smooth surface finish which did not present potential crack initiation points;

    (c) were free from crater cracks at their endings; and finally

    (d) possessed the required mechanical properties.

The specification for *Methane Princess* and *Methane Progress* was therefore drawn up in such a way that the builders were provided with very detailed guidance on how satisfactory results could be obtained – furthermore, the man who supervised and shared in the growing pains of the *Pioneer* tank construction was made freely available to both yards.

It was clearly emphasized that cleanliness, freedom from dust, rust, grease or excessive moisture, draughts, purity of the shield gas, were all essential ingredients for satisfactory welding. Care in sequence of work and weld preparation would minimize distortion. If welding could not be carried out under cover, in a clean shop, then adequate weather protection was essential, i.e. for site assembly on the berth. Careful quality control and tight management throughout the operation was vital.

Shipyard 'A' chose to ignore much of the proffered advice (it could not be imposed) and suffered – to its astonishment and cost – considerable difficulties and delays. Shipyard 'B' chose to heed the advice and not only met few problems but built the tanks within schedule and budget.

Needless to say, the experience of shipyard 'A' was the one which was

Plate 1. *Methane Progress* on trials.

Plate 2. Aluminium cargo tanks assembled on the berth at Harland & Wolff shipyard – *Methane Progress*.

broadcast to the world and established a reputation for aluminium welding which took some years to eradicate.

However, notwithstanding the trials and tribulations experienced during their construction, the *Princess/Progress* tanks remained in as-new condition during their lifetime – *Methane Progress*, 22 years, *Methane Princess*, 28 years and still in service. No leaks, no corrosion, no crater crack propagation, only a few early cracks at bottom bracket endings where built-in notches produced stress concentrations. These showed up in the first year of service, and the addition of local doublers to 'soften' the notch cured the trouble. Later abuse by hold flooding, warming up when locked in by ice, floating of two tanks and some buckling of internal structure, left them unaffected in their performance.

**Tank keying Design:** This presented a number of interesting problems at a time when three-dimensional finite element structural analysis techniques were not around to help:

> — how should the rolling forces be distributed between the top and bottom keys?
> — how much do the tanks and ship deflect when inclined? and what is the interrelationship between the two?
> — how much clearance, if any, should be allowed in the keyways?
> — how much would friction between the tank bottom and the balsa insulation contribute to the transverse loads at the bottom keys?

Unfortunately the *Methane Pioneer* provided no real clue as no instrumentation had been provided to measure loads at the keyway locations and the scale of the vessel was, in any case, rather too small to measure hull deformations in any meaningful way. In heavy weather, however, the top keys were inclined to make their presence known by a 'bumping' sound– which indicated some relative movement in this area.

As will be seen in other chapters, there were a large variety of ingenious solutions to the locating key problem but most were considered too complicated and, in the event, a simple male/female keyway–with careful assembly and minimum working tolerances–was selected, and this arrangement has given trouble free service.

In the absence of any better information ample design margins were incorporated, with 60 per cent of the total calculated transverse loads being provided for at the top keys, and 60 per cent at the bottom.

As to the bottom, a number of calculations and tests were carried out (a) on the balsa/hardwood panel keys and bonding between laminations; and (b) on the coefficient of friction between the aluminium and balsa. These

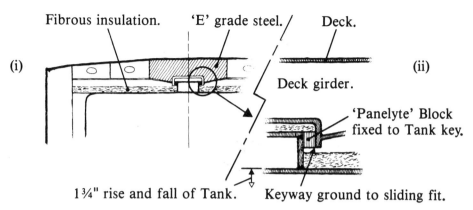

Fig. 9c. Keyways.

investigations culminated in the roughening of the tank bottoms to distribute as much of the transverse loads as possible over the bottom insulation system as a whole. By adopting such methods it was possible to incorporate very large factors of safety into the design.

**Vapour Venting Philosophy:** The design of the vapour relief system produced many arguments, not least with the authorities charged with their approval. The standard formulae of the time seemed unrealistic.

The major hazard to the ship seemed not so much to be the likelihood of every tank being totally enveloped in fire over its entire surface–this seemed quite unreasonable, though some degree of 'engulfment' certainly had to be designed for–but rather the effect of a tank overfill; this was always a nagging worry. Cost also came into the discussion.

In the event a common vent header was accepted, with relief valves of the simplest possible form (weight loaded) fitted at the base of two risers. This had the additional advantage of allowing accidental tank overflows to pass harmlessly into neighbouring tanks via the vent line.

The risk of damage to the common vent line by collision or the shipping of green seas seemed remote because LNG vessels by their nature have high freeboards and are therefore extremely dry ships; furthermore, the location of the header was at the ship's centreline and 8 feet above deck. Fig. 9d compares the above philosophy with present-day requirements; it has much to commend it.

(i)  Princess/Progress arrangement - Nine Tanks with two vent risers - relief valves at base of risers.

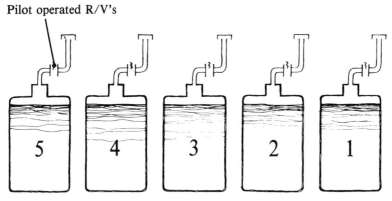

(ii)  Arrangement required by current International Regulations (relief valve(s) and riser at each Tank).

Fig. 9d.  Vent/Relief valve concept.

**Water Detection:** Part way through the construction programme it was decided that it would be prudent to make some provision for detecting leakage of water through the inner hull from the adjacent ballast tanks: it seemed that despite the great care applied to its design and assembly, and the application of top quality epoxy coatings, no-one was prepared to guarantee the integrity of the inner hull against minor leakage during the ship's contract life of fifteen years.

In line with the basic Conch philosophy of simplicity being best it was decided to incorporate a twin core electric cable behind the insulation at the aft end of each hold, terminating it in two bare wires at the lowest point;

this cable would be permanently connected to a warning lamp in the instrument room which would light up, red for danger, if seawater completed the circuit.

At the next plan review it was decided to terminate the wire at the top of each hold – but below deck, where it could be checked at each annual survey – thus saving the cost of about 500 feet of cable, plus warning lamps, etc. MISTAKE NO. 1.

After the first inner hull leakage, which occurred soon after both ships entered service, and with extremely inconvenient results, the wires were led back into the instrument room and checked at intervals. MISTAKE NO. 2. The intervals were not sufficiently frequent.

The next move was to fit more sophisticated equipment in the form of capacitance probes with a continuous readout in the form of green light for dry, red light for water. MISTAKE NO. 3. The equipment was not 'fail safe' and when the internals corroded and/or the probes were contaminated, the light still shone merrily 'green' under several feet of water!

MORAL: economy doesn't always pay off; and sophisticated equipment does not automatically provide the solution.

Inner hull leakage has caused several headaches and sleepless nights in these and other LNG ships, but is now reasonably well catered for.

**Remote Control of Cargo Handling Operations:** The *Princess/Progress* philosophy was simple: remote controls and instrumentation were not sufficiently reliable that they could be wholly relied upon; thus all significant day to day handling operations were manual or localized. The pump start/ stop buttons were located at each tank 'dome' – of course remote emergency stops were also provided; level gauges had local readouts *and* the tanks each had a viewing window as described earlier; relief valves were simple weight loaded devices; all liquid/vapour control valves were manual – except the ship side loading valves which had compressed air operators – locally controlled and fail safe.

Most important, the decks were flat and uncluttered and allowed the officer in charge to walk freely about his business of inspecting the performance of the equipment.

LNG has the convenient characteristic of manifesting its presence by frosting of uninsulated parts of the system, so that progress of cooldown and warm up of the piping system can easily be monitored by deck watchkeeping; leaking joints, harmless if detected at an early stage, also show up locally by a tell-tale white wisp of condensed water vapour.

Certainly these first ships were also quite small. Large ships and more

Plate 3. View of trunk top piping–*Methane Princess*.

reliable instrumentation and control systems have now resulted almost universally in cargo control rooms located either in the main accommodation aft deckhouse or on deck above the compressor house–not all LNG ships have means of convenient 'on-the-spot' inspection of the cargo system.

Discussions on the wisdom of almost total reliance on remote control of LNG handling operations will no doubt continue for many years to come.

**Dual fuel burning:** *Princess/Progress* were the first two ships to burn LNG cargo boil-off, along with bunker fuel in their boilers. Classification requirements demanded a 10 per cent pilot oil fuel at all times, and interlocks to prevent the existence of gas in the boiler before 'flashing up'; they also required double duct piping through which the gas was led inside the engine room. The system so developed caused few problems. The specially designed burners were tested in shore boilers well ahead of the ship delivery and, apart from some minor adjustments on trials to shorten the flame length, and some recurring problems in achieving a satisfactory automatic combustion control arrangement, the birth of dual fuel combustion was a painless one.

**Generally:** Based as they were on a philosophy of ample margins, simple design solutions wherever possible and the maximum use of proven technology, *Methane Princess* and *Methane Progress* provided a good foundation on which to build and develop.

In retrospect one of the major 'inconveniences' of the *Princess/Progress* design was the very small (approx. 3″) clearance around the cargo tanks – thus precluding internal inspection or, more important, the facility to repair the inner surfaces of the insulation. The extra cost in ship hull size was considered too great to justify incorporating a 18″ or 24″ hold space. For ships of 125,000m$^3$ capacity such spaces can be much more easily accommodated as for example in the Avondale ships subsequently designed by Conch and, more recently, the IHI design (see Chapter 16).

## JULES VERNE[5][6]

In France, though perhaps less well publicized at the time, work was progressing quite rapidly on the design of a ship to transport $50 \times 10^6$ scf/day of natural gas from Arzew to Le Havre.

The owners, la Société Gaz-Marine, had imposed several basic design and operating conditions which the selected shipyard, Ateliers et Chantiers de la Seine Maritime, was required to meet. These were:

— the ability to inspect visually both inside and outside surfaces of the cargo tanks;

— the capability of discharging the tanks, in an emergency such as pump failure, by applying pressure to the surface of the liquid;

— the tanks to be able to withstand the standard static water test as required by existing Classification Society rules;

— the tanks to be able to withstand an external overpressure;

— the tanks to be able to withstand any condition of partial loading, and the ship to have adequate stability under such conditions;

— the insulation to limit cargo evaporation to 0.27 per cent per day.

These were all in addition to meeting the current requirements of Bureau Veritas and maximizing the use of French manufactured materials and equipment.

Given the above design parameters the ship arrangement became as shown in figs. 9e and f–the tank design being strongly influenced by the requirement for pressure discharge.

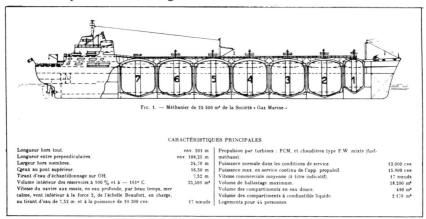

Fig. 1. — Méthanier de 25 500 m³ de la Société « Gaz Marine »

CARACTÉRISTIQUES PRINCIPALES

| | | |
|---|---|---|
| Longueur hors tout. | env. 201 m | Propulsion par turbines : FCM, et chaudières type F.W. mixte (fuel-méthane). |
| Longueur entre perpendiculaires | env. 188,25 m | |
| Largeur hors membres. | 24,70 m | Puissance normale dans les conditions de service. 13.000 cve |
| Creux au pont supérieur. | 16,50 m | Puissance max. en service continu de l'app. propulsif. 15.000 cve |
| Tirant d'eau d'échantillonnage sur O.H. | 7,52 m | Vitesse commerciale moyenne (à titre indicatif). 17 nœuds |
| Volume intérieur des réservoirs à 100 % et à — 161° C. | 25,500 m³ | Volume de ballastage maximum. 18.200 m³ |
| Vitesse du navire aux essais, en eau profonde, par beau temps, mer | | Volume des compartiments en eau douce. 490 m³ |
| calme, vent inférieur à la force 2, de l'échelle Beaufort, en charge, | | Volume des compartiments à combustible liquide. 2 150 m³ |
| au tirant d'eau de 7,52 m. et à la puissance de 10.300 cve. | 17 nœuds | Logements pour 44 personnes. |

Fig. 9e. *Jules Verne.*

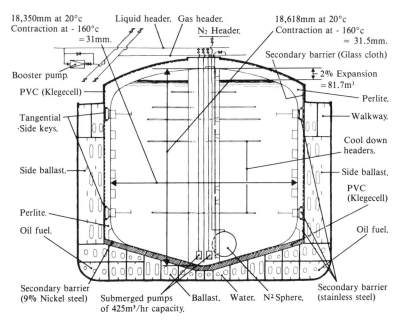

Fig. 9f.  Cross section of Cargo Tank.

Her principal particulars were as follows:

| | |
|---|---|
| Length overall | about 201.00 m |
| Length between perpendiculars | about 188.25 m |
| Breadth | about 24.70 m |
| Depth to upper deck | 16.50 m |
| Draft–scantling | 7.52 m |
| Tank volume (100 per cent full at −165°C) | 25,500 m³ |
| SHP (max continuous) | 15,000 cve |
| Speed (service) | 17 knots |
| Number of holds | 7 |
| Number of tanks | 7 |
| Material of tanks | 9 per cent Ni Steel |
| Max design stress (.375 × UTS [warm]) | 26.25 kg/mm² |
| Keys | at tank sides |
| Insulation | PVC foam and Perlite |
| Boil off (maximum) | 0.27 per cent/day |
| Overall discharge time | about 15 hours |
| Classification | Bureau Veritas |

Some additional background to the planning that went into the components of this ship is of interest:

**The tanks** were in fact a combination of a number of different geometrical shapes; the bottoms were, from the centre, part spherical, part conical, part elliptical-toroid; the sides were cylindrical; the tops were ellipsoidal. Situated at the central axis was a 3 metre diameter tube which housed all piping and gauging equipment, electric cables and access ladders.

The tanks were internally stiffened by two principal girders to resist 'ovalization' due to differential liquid heads in the inclined condition; these were supplemented by four secondary stiffening rings to resist external pressure.

Design stresses were confirmed by model tests to the following scales:

| | |
|---|---|
| model of the tank tops | 1:4.5 |
| model of the tank bottoms | 1:5 |
| model of the tank sides | 1:10 |

in which all relevant operating conditions were simulated.

The actual ship tanks were built up in the holds from pre-fabricated sections, with weather protecting covers over each hold to provide optimum conditions for welding and insulating.

Each tank was provided with lifting arrangements so that their undersides could be inspected during surveys – and so that the bottom insulation could be inspected and repaired. In fact, the tanks can be lifted 0.85 m without disturbing the top dome cover or the keys. Furthermore, in a real emergency they could be lifted right out of the ship without disturbing its structural integrity.

**Locating keys** were fitted at the sides (4 upper and 4 lower), in line with the primary internal ring girders – great care being taken to preserve the integrity of the secondary barrier at these locations. Each roll key was designed for 350 T; each pitch key for 100 T.

Stress distributions in the keys and tank walls were defined by the use of photoelastic modelling techniques.

**Insulation** was a combination of materials as already described and indicated in fig. 9f – the thickness of Perlite being determined by the space required for a man to inspect the outside of the tanks.

The PVC attached to the inner hull bottom was 45 cm thick; at the sides it was only 6 cm, augmented by 54 cm of Perlite.

The secondary barrier was also a combination of materials, depending on the location – the bottom was 9% Ni steel and in direct contact with the tanks; the sides were impregnated glass cloth, attached to the face of the PVC and designed with built-in flexibility; stainless steel was fitted around

Plate 4. *Jules Verne* cargo tank under construction.

Plate 5. *Jules Verne* under construction.

the keys to effect a liquid-tight barrier in these critical areas.

The designs for all details of the barrier were tested in liquid nitrogen or liquid methane either at the shipyard or the Gaz de France laboratories.

It will be seen that, in concept, the *Jules Verne* was perhaps even more cautious in its approach than the *Methane Princess* and *Methane Progress*; she entered service early in 1965 and has traded between Arzew and Le Havre since that date. In 1978 her fifteen year contract was extended for a further period of five years. She now (1992) trades between Algeria and Barcelona under new ownership, renamed *Cinderella,* and will have clocked up one million loaded miles in service and discharged over 900 cargoes by the end of 1993. A great achievement by any standard.

## OPERATING EXPERIENCE
So, with the benefit of hindsight, how did they perform? On balance, all three ships have had an excellent service record.

**LNG/LN$_2$ spillages:** Despite carefully written operating manuals and laid down procedures, all three ships have experienced spillage of either LNG or liquid nitrogen on deck.

*Methane Princess* and *Methane Progress* have cracked their decks, not seriously, because of
   (1) not observing the need to drain completely the deck lines before disconnecting the loading arms – the valve was leaking and over-filled the (too small) drip pan.
   *Result*: cracked deck extending about 4 ft.
   (2) lack of watchfulness while topping up the liquid nitrogen storage tanks, causing an overflow onto the deck.
   *Result:* cracked foredeck, and deck below, extending over about 100 sqft.
   (3) over-keen topping up of tanks, causing liquid carry-over into vent header and spray of liquid out of riser.
   *Result*: nothing but slight shock to personnel.
*Jules Verne* experienced overflow of the forward tank in the first year of operation due to faulty instrumentation. (The float gauge had failed, and some of the tank wall thermocouples were inoperative.)
   *Result*: liquid spill from vent header onto the dome cover which cracked this and propagated into the upper deck, which may also have been partially embrittled.

The out of service time was, except in the case of *Jules Verne*, small or non-existent. The latter spent a few days at the repair yard.

**Inner hull leakage:** *Methane Princess* and *Methane Progress* have both experienced a number of inner hull leakages into the space behind the insulation – in two instances building up sufficient pressure for the water to break through the insulation into the holds. This is a very inconvenient situation for this design, since there is no space between the tanks and the insulation, and all repairs have to be made from the ballast tanks. Surprisingly, however, and this was confirmed by tests, wetting of the balsa wood has little effect on its insulating properties – or its coefficient of friction against the tanks, although some temporary swelling of the panels is inevitable.

The *Jules Verne*, but much later, experienced a similar incident, likewise inconvenient and necessitating a spell in drydock for structural modifications; it did, however, provide an opportunity to examine the PVC below the tanks which, after mechanical testing, indicated that no physical ageing – in the form of fall-off in mechanical properties – had occurred after approximately ten years in service.

**Cargo pumps** were originally fitted with some apprehension as to their reliability and for this reason were provided with back up systems to safeguard against unforeseen pump failure. These units have however remained very trouble-free – unless they are provided with a diet of too much rust, scale, sand or other debris, in which case they are rather understandably liable to protest.

In summary, therefore, it can be said that most of the problems experienced by the early – and even later – LNG ships to date have been due to the conventional parts of the ship developing faults or to careless operation – i.e. the human factor.

[1] 'Low-temperature, liquefied-gas transportation' by C. G. Filstead and Montgomery Banister. SNAME, Vol 69, 1961.

[2] 'The UK liquid methane tankers' by C. G. Filstead and D. E. Rooke. Petroleum and the Sea Conference, Sect III No 304, 1965.

[3] 'Six years operational experience with the *Methane Princess* and *Methane Progress*' by C. G. Filstead and D. E. Rooke. LNG 2, Paris, 1970.

[4] 'Twelve years operating experience with *Methane Princess* and *Methane Progress*' by P. L. L. Vrancken and J. McHugh. LNG 5, Dusseldorf, 1977.

[5] 'Le navire méthanier *Jules Verne*' by J. Grilliat. Association Technique Maritime et Aéronautique, 1964.

[6] 'Le méthanier *Jules Verne*' by A. Gilles. Petroleum and the Sea Conference, Sect III No 303, 1965.

# 10
# Patents and Politics

'What we claim is . . .': the now familiar opening to the claims section of the patent is, in retrospect, a not altogether inappropriate description of the political influences which have also influenced the LNG scene.

## PATENTS . . .
What is a patent and what is its purpose? This question was put to one of the patent lawyers who figured prominently in LNG affairs; his reply was as follows:

> 'The fundamental idea behind every patent law is the same – a grant of exclusivity by the State for a specified term of years as a reward for making an invention and for disclosing it to the public.'[1]

Its purpose? – 'to encourage invention'.

The first recorded grant of exclusive rights to make and sell articles was contained in a law promulgated by the Venetian State in 1474; which required that

> 'whoever made any new and ingenious manufacture should register it with the city authorities and that it would be forbidden to anyone else in the land to make any other artifice to the image and similarity of that one without consent and licence of the author during the term of ten years.'

From the start, the Constock company adopted a policy of worldwide patent coverage not only to protect their research and development work so as to ensure that some financial return was obtained from the considerable expenditure involved, but also to ensure that their early technological lead was maintained. They were not alone in adopting this attitude; indeed the patent literature of the ten year period from the mid-fifties to mid-sixties is quite liberally sprinkled with original ideas for the carriage of liquefied gas by sea.

In 'searching for a solution', many ideas were produced, many unquestionably novel or inventive, but relatively few either merited, or were capable of, practical application.

There is no doubt that at this time the shipbuilding industry was not patent conscious; unlike other industries such as oil, chemicals and pharmaceuticals, very few of the firms involved paid much attention to the patent system and Constock's new approach hit them hard—they couldn't really understand it. Patents are a mystery and an irritation to the Engineer—or Naval Architect— call him what you will, whose terms of reference are quite simply to solve a given technological problem in a safe, reliable and cost effective manner.

When is an idea inventive—and therefore patentable, and when is it a normal piece of engineering development by 'one practised in the art'? What is the difference between a straightforward patent, a combination patent and a patent of improvement? Does one 'infringe' or claim 'non-use'? Is the prior art significant? What is the public domain? Is a patent strong or weak? This was the unfamiliar, and largely uncharted, territory in which shipowners, shipbuilders, Classification Societies, independent design groups and their employees all found themselves entangled from the moment they ventured—or were drawn—into the LNG web.

Patent Agents, and Attorneys, will provide immediate answers to all these questions and more; will write comprehensive patents with the widest possible claims, embracing 'solutions' which the engineer is quite convinced have been in use for at least 100 years; will explain that the patent office of each country takes a different attitude to patentability—or novelty—which we *think* is the same thing; will point out that the duration of validity varies— as does the cost of maintaining validity; will explain that the 'examination' process, publication date and date of grant also differ so widely that in one case, Belgium, publication date follows the application date in a matter of weeks whilst in others the date of publication tends to coincide with the expiry of validity. To add to the confusion, the language of patents, in trying to be precise, can produce some very obscure descriptive words which require a classical education to understand—thus providing an additional hazard for the engineer.

It is a field into which the engineer is recommended not to venture, being at the same time fascinating and frustrating, profitable and profitless, time consuming and mysterious to the bitter end; it is the happy hunting ground of the devious minds of commercio-politicians. The engineer is lost there— far better for him to look ahead and to concentrate on solving his own problems.

A review of patent literature does, however, provide an interesting insight into the ways in which designers have sought to solve some of the problems which are fundamental to LNG containment systems. It will be seen that a number of these solutions were, in fact, 'too early', there being no practical means of implementing them at the time, but as analytical methods improved (with the introduction of computers) and methods of application advanced (with new materials and welding techniques), many ideas became viable. Many more were 'non starters' from their inception—and will forever so remain.

A number of these 'inventions' are illustrated and briefly discussed below; they appear in approximately chronological order of filing date and are divided roughly into the subject headings of (a) independent tanks, (b) tank supports, (c) integral tanks, (d) membrane tanks, (e) insulation systems, (f) miscellaneous; there is, inevitably, considerable overlapping between the five main categories. Furthermore, the chapter tends to dwell on earlier patents rather than those developed in more recent years.

**Independent Tank Systems**
The earliest patents were necessarily of a conceptual character and described the liquefaction and transportation in very general terms. For instance, Morrison's 'Method of Liquefying Gas' filed in 1951 (fig. 10a) was concerned with '. . . improvements in apparatus of storing and shipping . . . methane . . . which has heretofore been transmitted in pipelines'.

The patent goes on to describe the 'apparatus' which

> '. . . involves a compressing and refrigerating system at a point of origin, a peripatetic, or wandering, insulated receiver and a place to put the gas at the end of the line . . .'.

Note the language: the 'insulated receiver' is the ship, or barge; 'peripatetic' is of Aristotlean derivation, from the philosopher's habit of walking in Lyceum while teaching!

A later Morrison patent of May 1954 provided more details of the barge although the claims were primarily concerned with the liquefaction process (fig. 10c); by now the barges had developed five individual tanks, of 7,000bbls

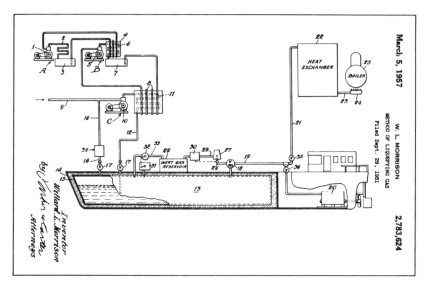

Fig. 10a.

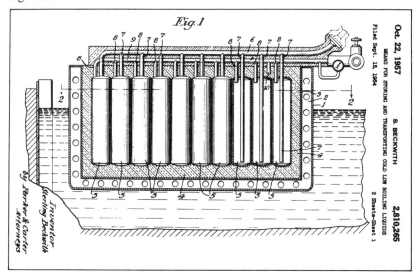

Fig. 10b.

capacity each, a double hull construction, whereby the space between hulls can be flooded to 'control draft and constitute an additional safety factor in the event of collisions'.

The same year (September 1954) Beckwith, one of the Constock project engineers, produced a design for a multi-vessel containment system (fig. 10b; liquid sloshing was even then becoming a matter of some concern because the descriptive matter states that

'... if large tanks are to be shipped in a barge or ocean-going vessel, then swash plates of substantial size and strength and complication must be inserted in the

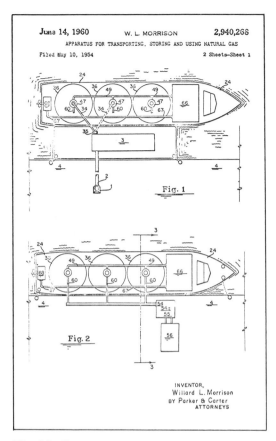

Fig. 10c.(i)

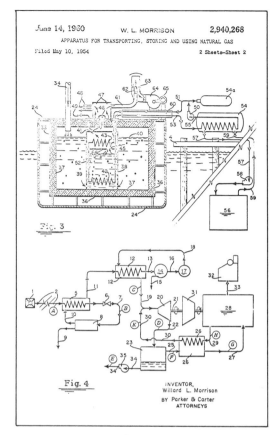

Fig. 10c.(ii)

tank, in contact with the lining, thus adding greatly to the expense and complication of the structure because large volumes of liquid when violently agitated, generate in the tank very high presures of impact. . . .'.

For this reason Beckwith proposed a large number of independent tanks in an insulated hold.

One year later, in December 1955, both Howard of Esso Research & Engineering, and Rupp of the same company, produced designs for independent tanks (figs. 10d & 10e). Howard's idea was interesting in that it comprised a double walled tank incorporating insulation between the two walls. Around the outside wall, and in direct contact with it, is carried a liquid oil product '. . . capable of absorbing substantial quantities of escaping gases . . .'; an ingenious method of dealing with LNG leakage.

Rupp's design envisaged a secondary container within the insulated holds filled, or part filled, with $CO_2$ which, when solidified, provided support for the LNG tank.

It is interesting to observe that the double-shell tank concept, first developed

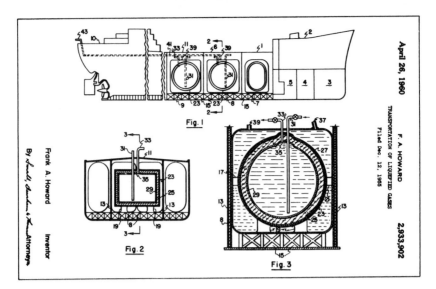

April 26, 1960    F. A. HOWARD    2,933,902

TRANSPORTATION OF LIQUEFIED GASES

Filed Dec. 12, 1955

Frank A. Howard

Inventor

By ⟨signature⟩ Attorneys

Fig. 1

Fig. 2

Fig. 3

Fig. 10d.

by Esso in 1955, seemed to persist in their basic thinking for well over ten years until it finally emerged as a truly double walled independent tank design, which was incorporated into their four 40,000 m³ ships built for the Libya/Italy/Spain project (fig. 12b).

By the end of 1955 the J. J. Henry Company, now retained by Constock as Marine Consultants, were beginning to crystallize their ideas on an ocean-going ship design. Jim Henry's patent of May 1956 described a general ship layout, paying specific attention to the way in which the tank domes might be flexibly connected to the deck (to allow for relative expansion and contraction of the tanks), means for circulating 'warm' ballast water within the ballast tanks, and describing how the basic concept could be applied to a T.2 tanker (the first proposal for a prototype). See figs. 10f(i)-(iv).

This was followed a year later, in August 1957, by a 'patent of improvement' which defined more precisely the way in which the tanks were housed 'in a manner to maintain the tank in predetermined relative position while permitting relative expansion and contraction movements' and the way in which heat could be applied to the inner hull plating for temperature control (fig. 10g).

Of this patent C.I. Kelly wrote in the *Petroleum Times* of 15 January 1960:

> '[It] became public on July 28th, 1959, two months after Constock's paper had been read in New York . . . to J. J. Henry . . . the designer of the *Methane Pioneer*. Those who have seen and observed the ship will be forgiven perhaps for believing the similarity between the paperwork and the actual object is more than accidental, and that the paperwork is aimed at protecting designs of future larger editions of the *Methane Pioneer*.'

Mr. Kelly clearly did not approve of patents!

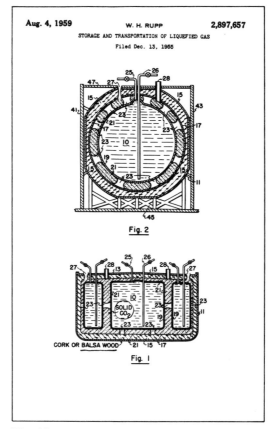

Fig. 10e.

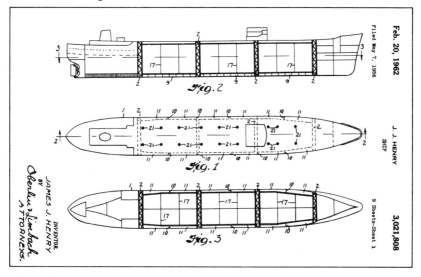

Fig. 10f.(i)

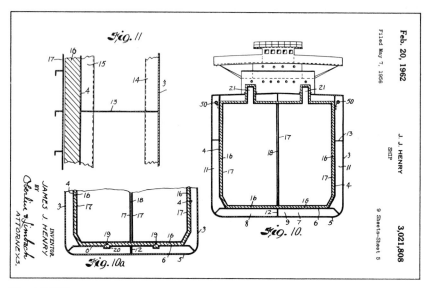

Fig. 10f.(ii)

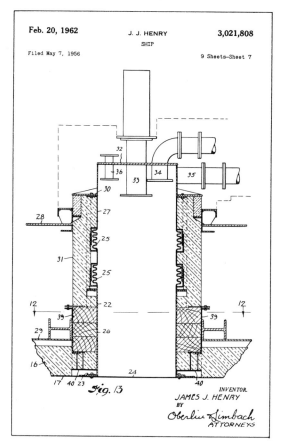

Fig. 10f.(iii)

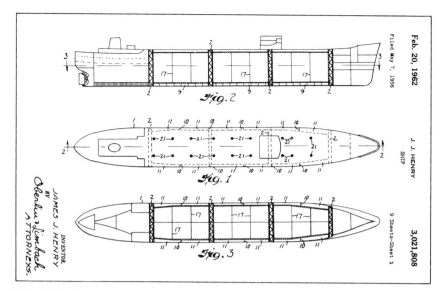

Fig. 10f.(iv)

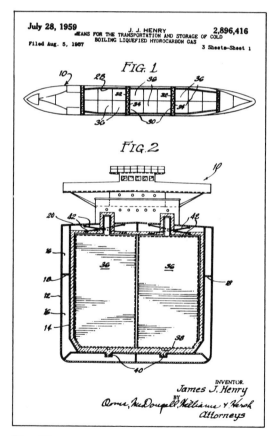

Fig. 10g.

103

There is no doubt that it was this patent, applied for and granted worldwide, which so surprised everyone by its 'audacity', and which so strongly influenced subsequent LNG ship designs; it described what was to become the Conch Independent tank system and—but later—the 'conventional' method of building LNG tankers. It covered, as can be seen, a double hulled ship, divided into several holds each insulated on the internal surfaces; within the insulated holds were fitted one or more independent tanks which were free to expand and contract.

Most particularly this patent surprised naval architects and shipbuilders, for how else could an LNG ship be designed? And by what stretch of the imagination was a double hulled ship patentable? After all, every naval architect knew that the *Great Eastern* had a double hull—and that was a hundred years ago—and, anyway, a double hull was a classification requirement for this type of ship. Of course we now know that the trick is to produce a 'combination patent'!

However, naval architects by nature and training are nothing if not ingenious, and they soon began to come to grips with the mysteries of Patents and, what was much more important, discovered the chinks in their armour—for chinks there surely were.

In that same year (1959) Burness, Kendall and Partners, under their assignment from William Cory & Son (Chapter 4), had patented their own design with cylindrical tanks (fig. 10h); this proposal also incorporated means of flooding the spaces around the cargo tanks with 'water, or other liquid such as oil, which will freeze at a temperature of methane…and effectively plug any leaks from the methane tanks'; this liquid was to be stored in deck tanks.

The reason for the 'tumble-home' of the ship's sides is not explained.

Also in the same year Hans Lorentzen patented his spherical design (fig. 10i[i]) with further improvements in 1960 (fig. 10i[ii]).

Gebien's patent of 1959 (fig. 10j) provides solutions (a) to the tank support problem—by floating it; (b) to the leakage problem—by providing a 'self sealing' tank and (c) to the expansion/contraction problem—by 'incorporating a non-metallic receptacle which is not affected by the thermal expansion…' —simple!

Esso Research and Engineering did not finalize their design for a double walled LNG tank system until 1966 and then only after considerable efforts to avoid Conch's wide ranging portfolio; note the side keys and insulation attached to the tank (fig. 10k) – both so located very largely for this reason.

**Tank Supports**

The problem of how to support a tank securely in a ship without imposing

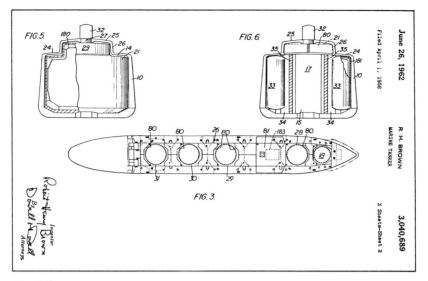

Fig. 10h.

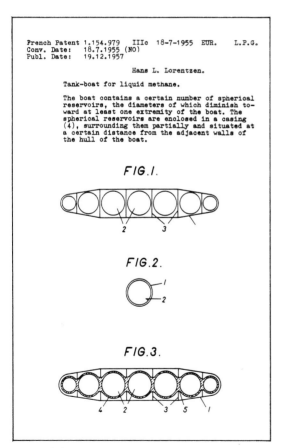

Fig. 10i.(i)

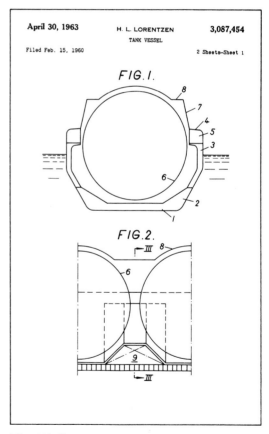

Fig. 10i.(ii)

105

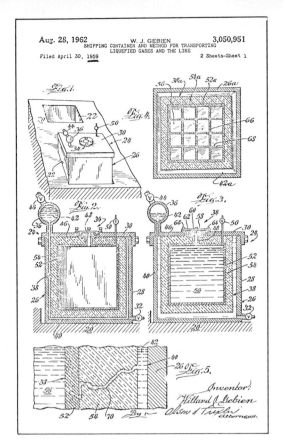

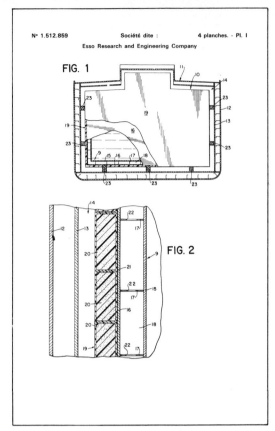

Fig. 10j.                                          Fig. 10k.

undesirable stress concentrations either in the tank or ship was a very real
one and there were many potential 'solutions' as the patent literature testifies;
a selection is illustrated below, by no means comprehensive. But the best is
often the simplest and there is little doubt that the Conch and IHI (trapezoidal tanks) and Moss (spherical tanks) win on this score.

**Integral and semi-membrane designs**
A number of designers sought to eliminate the problems presented by the
expansion and contraction of independent tanks, and the influence of ship
structural deflections, by incorporating the tanks as part of the ship's structure
in some way or another.

The basic problem which then arose was how to analyse the combination
sufficiently accurately to be confident of success, and it was this fact which
prevented the development of the concept until Kvaerner Moss produced
their sphere support and cover design with computer aided techniques. There
were many heated arguments within the Shell organization as to the practi-
cability of analysing the 1956 BPM design (fig. 10m[i]).

106

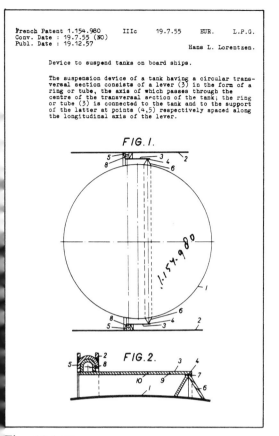

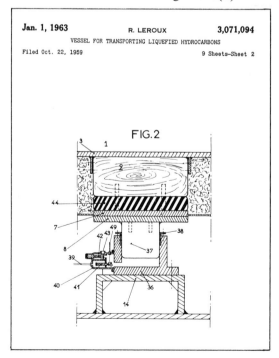

Fig. 10 l.(i)

Fig. 10 l.(ii)

Fig. 10 l.(iii)

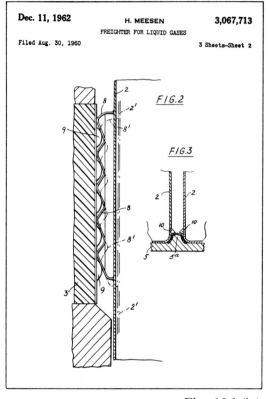

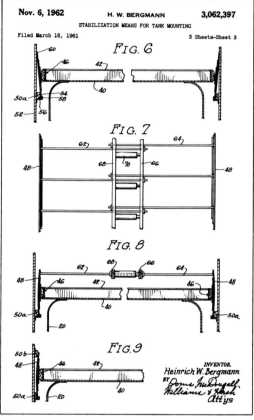

Fig. 10 l.(iv)

Fig. 10 l.(v)

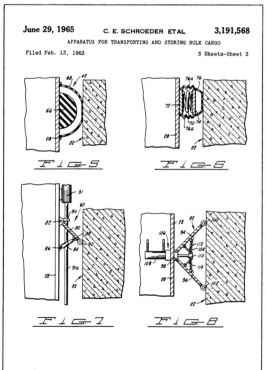

Fig. 10 l.(vi)

Note: See also Chapter 14 for methods devised for supporting spheres.

Some of the early patents are shown in fig. 10(m).

The Stowers (Texas Gas) patent envisaged filling the space around the tank with a liquid which solidified at −165°C, increasing the level progressively as the cargo was loaded. The tank was finally held firm by the solidified product.

M. Leroux (Dubigeon Shipyard, France) had a very fertile mind and produced a number of interesting design concepts.

Dr. Yamamoto's semi-membrane was the first original LNG design to emanate from Japan. It has subsequently been successfully adopted for fully refrigerated LPG ships which operate at temperatures of about −45°C, but not for LNG.

**Membranes**

Starting from 1954 when Morrison found that it would be necessary to line his internally balsa-insulated tanks with a liquid-tight metallic lining, there were many attempts to find a satisfactory solution to this problem. A few of these attempts are illustrated here, see figs. 10n.(i) to (xvi), amongst which may be recognized the origins of the present systems.

Many were rejected, but as many were developed to an advanced stage−in terms of manufacture and prototype testing−only to be found lacking in the essential resistance to long-term fatigue.

The next chapter discusses the development of the successful membrane designs in more detail.

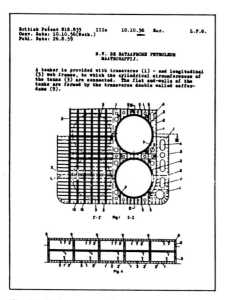

Fig. 10m.(i) Shell-integrated tank design.

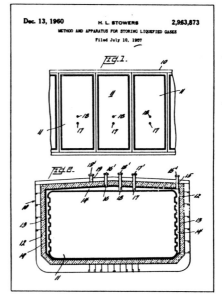

Fig. 10m.(ii) Stowers–semi-membrane design.

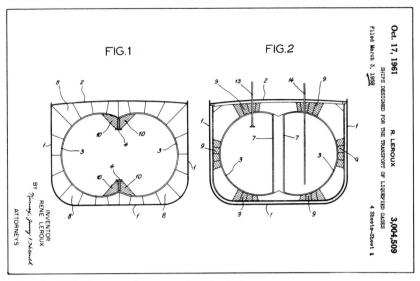

Fig. 10m.(iii)  Leroux – integrated tank design.

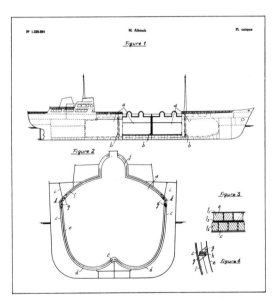

Fig. 10m.(iv)  Albiach – semi-membrane design.

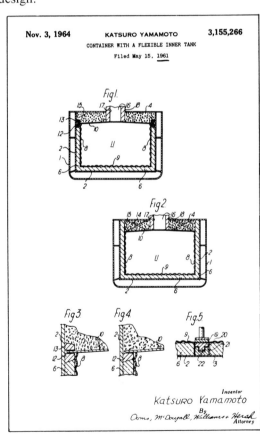

Fig. 10m.(v)  Yamamoto –
semi-membrane design.

110

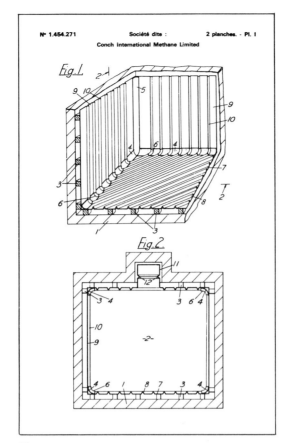

N° 1.454.271      Société dite :      2 planches. - Pl. I

Conch International Methane Limited

Fig. 10m.(vi)  Conch–semi-membrane design.

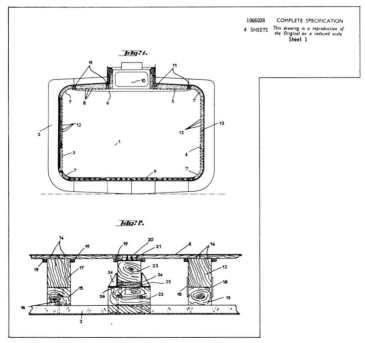

1066039      COMPLETE SPECIFICATION

4 SHEETS    This drawing is a reproduction of
the Original on a reduced scale
Sheet 1

Fig. 10m.(vii)  Atlantique–semi-membrane design.

111

## Insulation

Although presenting many technical problems the patents relating to insulation *per se* are relatively few—the insulation generally appearing as a component part of a total system. Balsa wood, and its assembly into prefabricated panels, dominated in the early years, confirming the early designers' confidence in the material, but its cost of fabrication eventually caused its demise in favour of polyurethane foams.

As an alternative, Polyvinyl and Polyurethane foams in their many forms and manifestations, now tend to lead the field; the assembly of the preformed blocks in such a manner as to prevent heat leaks at the joints has been the main concern of patents in this area. See Figs. 10.o. Page 117.

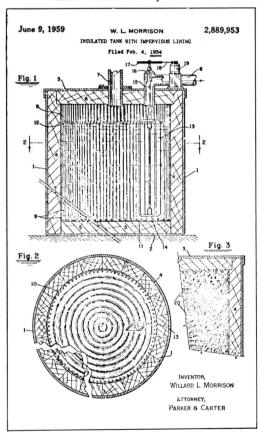

Fig. 10n.(i)

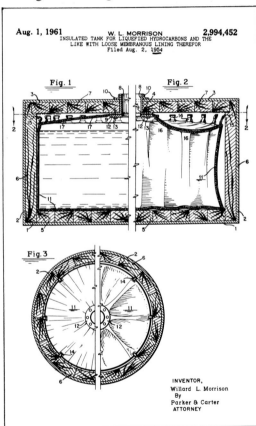

Fig. 10n.(ii)

Fig. 10n.(i-xvi) Various membrane proposals including the original MORRISON design (i).

112

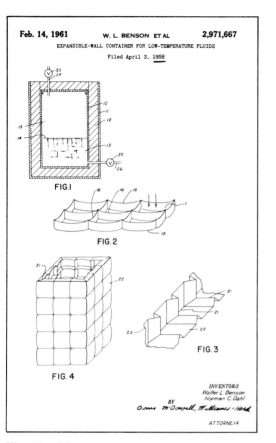

Fig. 10n.(iii)

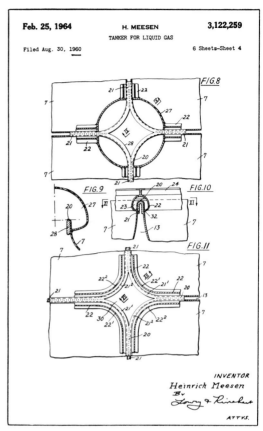

Fig. 10n.(iv)

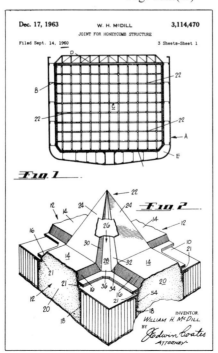

Fig. 10n.(v)

113

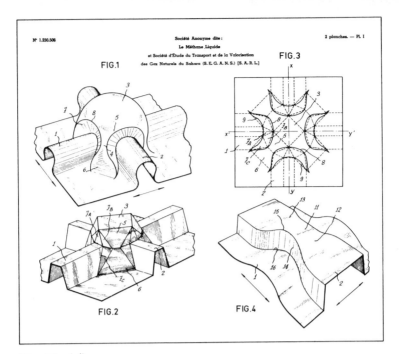

Fig. 10n.(vi)

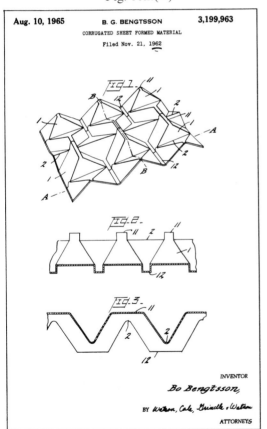

Fig. 10n.(vii)  Bengtsson's membrane
(later acquired by Technigaz.)

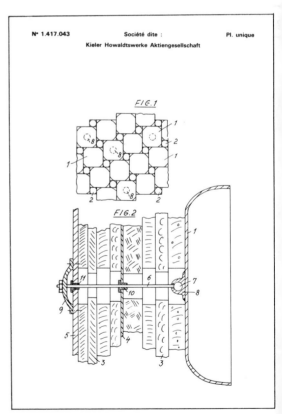

Fig. 10n.(viii)  Kieler Howaldtswerke membrane

Neth. Appl. 6.504.207    IIIc    2.4.64    Eur.    PG
Conv. Date: 2.4.64 (France)
Publ. Date: 4.10.65

SA de GÉRANCE et d'ARMEMENT.

Semi-rigid sheet material that is able to withstand extreme
temperature variations, e.g. as the lining of cryogenic
storage tanks in ships' holds, is divided into square or
rectangular flat portions by a pattern of V-shaped compensation corrugations (2). At the points where two such corrugations intersect, the thermal movement of the corrugations themselves is taken up by a pyramidal depression (8)
with an octagonal ground plan. The sides of the depression
are formed by folds extending from the crests (5) of the
compensation corrugations to the points of intersection
of the base lines (6,7) of those corrugations.

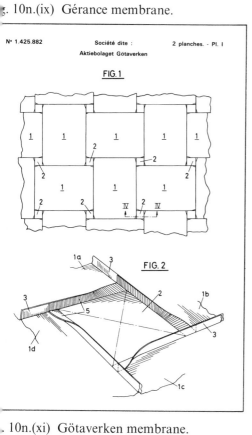

Fig. 10n.(ix)  Gérance membrane.

N° 1.425.882    Société dite :    2 planches. - Pl. I
Aktiebolaget Götaverken

### FIG.1

### FIG. 2

Fig. 10n.(xi)  Götaverken membrane.

N° 1.387.955    Société dite : Technigaz    Pl. unique

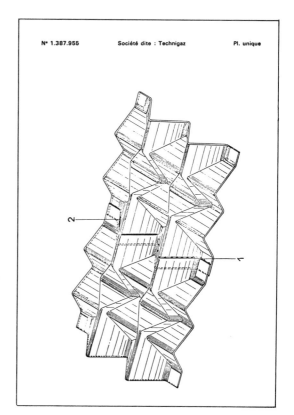

Fig. 10n.(x)  The basic Technigaz design.

French Patent 1.438.330    IIIc    3.3.65    Eur.    PG
Appl. Date: 5.3.65
Publ. Date: 13.5.66

GAZ-TRANSPORT.

Insulated tanks for liquefied natural gas are constructed
in double-walled ships holds in the form of two concentrical,
fluid-tight, semi-rigid shells (37 & 8) which are spaced
from each other and from the hold (3,4), with insulating
slabs (1,14) filling the intervening spaces. Those slabs
are e.g. wooden boxes filled with insulating material, e.g.
perlite. They are secured to the wall of the hold by means
of holding bars (2,29), the bars (29) for the inner layer
of slabs (14) being mounted by means of wooden cleats (15)
secured (16) to the hold wall and supporting the outer semi-
rigid shell (8). The semi-rigid shells (37 & 8) are composed
of steel sheets having a high nickel content. The sheets
have inwardly projecting flanges which are welded together
with the interposition of the web of a corner bar (10,38) or
the like, by means of which bars the shells are secured to
the underlying insulating slabs.

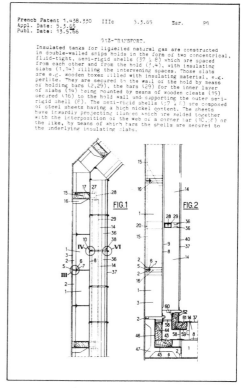

Fig. 10n.(xii)  The basic Gaz transport design.    115

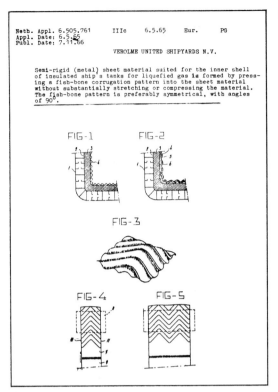

Neth. Appl. 6.505.761    IIIc    6.5.65    Eur.    PG
Appl. Date: 6.5.65
Publ. Date: 7.11.66

VEROLME UNITED SHIPYARDS N.V.

Semi-rigid (metal) sheet material suited for the inner shell
of insulated ship's tanks for liquefied gas is formed by press-
ing a fish-bone corrugation pattern into the sheet material
without substantially stretching or compressing the material.
The fish-bone pattern is preferably symmetrical, with angles
of 90°.

Fig. 10n.(xiii)   VEROLME membrane.

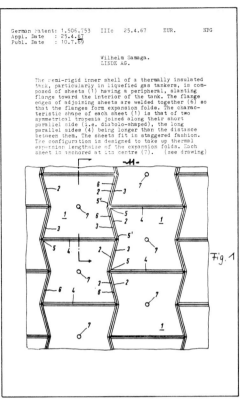

German Patent: 1.506.753    IIIc    25.4.67    EUR.    NPG
Appl. Date   : 25.4.67
Publ. Date   : 10.7.69

Wilhelm Samaga.
LINDE AG.

The semi-rigid inner shell of a thermally insulated
tank, particularly in liquefied gas tankers, is composed of sheets (1) having a peripheral, slanting
flange toward the interior of the tank. The flange
edges of adjoining sheets are welded together (6) so
that the flanges form expansion folds. The characteristic shape of each sheet (1) is that of two
symmetrical trapezia joined along their short
parallel side (i.e. diabolo-shaped), the long
parallel sides (4) being longer than the distance
between them. The sheets fit in staggered fashion.
The configuration is designed to take up thermal
expansion lengthwise of the expansion folds. Each
sheet is anchored at its centre (7).   (see drawing)

Fig. 10n.(xiv)   LINDE membrane.

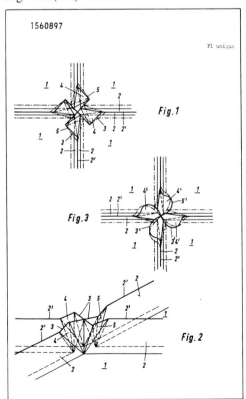

Fig. 10n.(xv)   LINDE membrane (cont.).

116

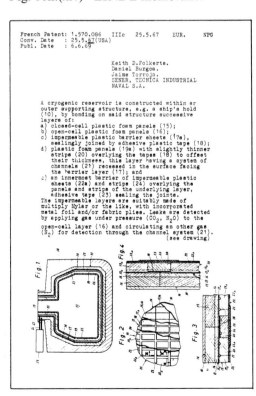

French Patent: 1.570.086    IIIc    25.5.67    EUR.    NPG
Conv. Date   : 25.5.67(USA)
Publ. Date   : 6.6.69

Keith D.Folkerts.
Daniel Burgoa.
Jaime Torroja.
SENER, TECNICA INDUSTRIAL
NAVAL S.A.

A cryogenic reservoir is constructed within an
outer supporting structure, e.g. a ship's hold
(10), by bonding on said structure successive
layers of:
a) closed-cell plastic foam panels (15);
b) open-cell plastic foam panels (16);
c) impermeable plastic barrier sheets (17a),
   sealingly joined by adhesive plastic tape (18);
d) plastic foam panels (19a) with slightly thinner
   strips (20) overlying the tapes (18) to offset
   their thickness, this layer having a system of
   channels (21) recessed in the surface facing
   the barrier layer (17); and
e) an innermost barrier of impermeable plastic
   sheets (22a) and strips (24) overlying the
   panels and strips of the underlying layer,
   adhesive tape (23) sealing the joints.
The impermeable layers are suitably made of
multiply Mylar or the like, with incorporated
metal foil and/or fabric plies. Leaks are detected
by applying gas under pressure ($CO_2$, $N_2O$) to the
open-cell layer (16) and circulating an other gas
($N_2$) for detection through the channel system (21).
                            (see drawing)

Fig. 10n.(xvi)   SENER membrane.

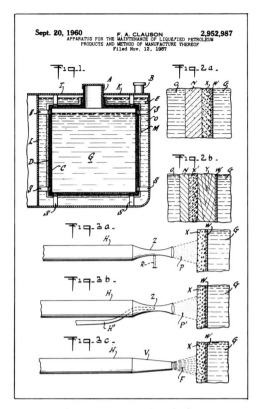

Fig. 10o.(i)  The Texaco insulation system.

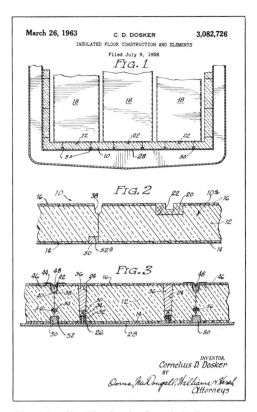

Fig. 10o.(ii)  The basic Conch Balsa panel system.

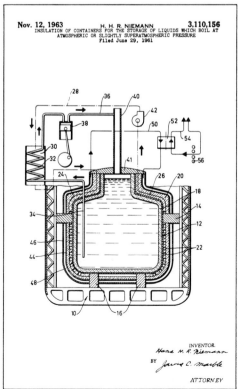

Fig. 10o.(iii)  The Minikay design.

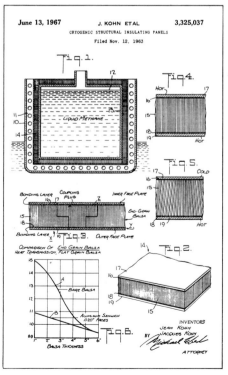

Fig. 10o.(iv)  A later Kohn (Belco    117 Balsa) design.

**General**

The selection of patents illustrated in this chapter, represents a very small collection of the total literature; at one time Conch alone held a portfolio of 1,500 patents worldwide, of which 500 were exclusively concerned with marine transportation. A portfolio of this size provided a powerful bargaining position but, as we know, could not prevent competitive systems being developed and successfully marketed.

Establishing a patent position by definition obliges a company to defend that position and Conch defended its technology vigorously as each suspected infringement occurred. Technigaz were the first victim with their membrane prototype *Pythagore*; Esso next with their double walled independent tank; Gaz Transport third with their Invar membrane design; somewhere in the middle J. C. Carter with their retractable pump, and finally the Kvaerner Moss design. Indeed, Conch were able to claim in the mid 1970s that not a single containment system in commercial service had completely escaped its net and that each paid it a royalty of some kind.

Not long afterwards the company left the scene in a rather dramatic manner – but that had nothing to do with patents as will be seen in a later chapter.

Conch were not the only group to vigorously defend their patents, witness Kvaerner v. Sener (Chapter 13).

It has always been one of the shipbuilders' greatest concerns to ensure that, in accepting a licence to build an LNG ship, they are adequately indemnified by the licensor against patent infringement. As already mentioned, the Classification Societies also became entangled in the net, finding themselves in an embarrassing position when attempting to provide guidance to shipyards and design groups. Recommendations as to certain arrangements or design features which they felt would result in an improvement or be more acceptable, often led to the risk of patent infringement, since it will be recalled that their rules were by no means specific at this time. Thus the Classification Societies themselves, always the staunch guardians of commercially confidential material, were now also required to familiarize themselves with a broad understanding of the patent situation and on occasions were inhibited from providing solutions which, in other circumstances, they would have considered their duty to pass on for the benefit of the industry.

In summary therefore, the patent policy of the early LNG design groups may be said to have
- largely failed, in that its objective to guarantee a technological lead to any single group was not achieved;
- partially succeeded, in that it provided a licence income with which to fund further development work;

- stimulated efforts by others to devise alternative methods of transporting LNG;
- created an extraordinarily diverse and, in many cases, ingenious, variety of solutions to a single set of design problems; which
- by its very diversity, probably stimulated the early introduction of an international code which in turn brought compensatory benefits to the industry.

It is for debate whether, on balance, a policy of establishing and maintaining a large patent portfolio, with its attendant expense in money and manpower, is worthwhile. The reader may form his own opinion; the author's may be apparent by reading between the lines of this chapter.

## ... AND POLITICS

If patents created problems, politics only served to compound them. While company politics are a normal engineering hazard, and relatively well understood, national and international politics are another matter and these began to interfere at an early date.

One of the conditions imposed on the first LNG project was that the ships should be constructed in the UK. Similarly the ship required for deliveries to France should be built there and, furthermore, should incorporate the maximum content of French manufactured components.

The construction of the four Esso Libya/Italy/Spain vessels was 'allocated' to Italian and Spanish yards in precisely the proportion of the deliveries to those two countries; similarly *Methania* was built for the Algeria-Belgium service in 1982 by the Boelwerf yard in Belgium.

The Alaska/Japan ships, by contrast, were allocated to the most competitive shipyard – Kockums in Sweden, and the seven 75,000m$^3$ vessels required for the Shell Brunei-Japan project were built in France with the benefit of substantial subsidies from both the British and French governments of the day. Not long afterwards both the Indonesia-Japan and El Paso Algeria-U.S.A. projects were both based on ships built in the U.S.A. and sailing under the U.S. flag, duly benefiting from useful U.S. subsidies.

By this time Japan, by far the largest gas importer, had become tired of seeing these lucrative ship construction contracts going to 'foreign' yards and imposed a requirement that a substantial percentage of the ships required for future LNG imports should be built in Japanese yards – and, as the expertise built up, operated by Japanese owners.

It is not the purpose of this book to dwell on politics, a fruitless exercise, but merely to draw attention to its existence as an important influencing factor in technological development.

119

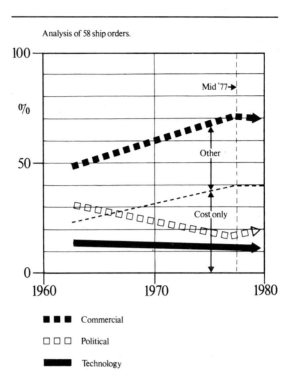

Fig. 10p. Factors influencing allocation of ship orders.[2]

[1] 'Patents' by J. M. Soesan. *Chemical Engineering Practice*, Vol II, Sect 9, 1959.

[2] 'A review of current LNG ship technology − and an attempt to rationalise' by R.C. Ffooks. LNG 5 Conference, Düsseldorf 1977.

# 11
# Membranes – a Breakthrough?

The membrane system heralded a much sought after cost breakthrough in LNG technology and, at the same time, introduced a totally new design and construction concept to the shipbuilding industry.

The first recorded ideas for membrane designs were patented by Constock in 1954, but there is no record of their having been seriously developed; they were, in reality, attempts to find a method of preventing LNG from penetrating, and damaging, the balsa linings installed in the early barge storage units. However, there was no doubt that if it were possible to devise an effective liquid tight lining for a load bearing insulation system, there were considerable potential cost savings to be made – or so it seemed, because

— the ship structure itself could absorb the cargo load;
— the high cost of special cryogenic steels or aluminium for separate tanks could be reduced by about one twentieth, and
— the space around the independent tanks could be eliminated, thus improving the ship's volumetric efficiency.

However, to achieve this desirable result without prejudicing safety margins – an essential prerequisite of any design change – it was necessary to develop a system in which

— the membrane lining should be able safely to withstand all operational temperature changes and gradients;

— the insulation should be able to withstand the imposed hydrostatic and dynamic liquid loads;

— a secondary barrier should be incorporated within the system;

— the system as a whole should be suitable for assembly using as near normal shipyard techniques as possible;

— the arrangement should be amenable to regular inspection and repairs which, in turn, should not cause undue delays to the ship;

— it should be possible to accept ballast water leakage through the inner hull without damage;

— the membrane should be readily weldable, preferably by semi- or fully-automatic methods if at all possible;

— the minimum extent of special construction equipment or operational restrictions should be imposed;

— the membrane, being thin, should be resistant to corrosion under normal LNG ship operating conditions;

— ship structural deflections, both local and 'global', would be transmitted directly to the membrane via the supporting insulation system and these must be safely absorbed.

To meet these basic design requirements was a tall order, but a number of proposals were produced of greater or lesser complexity and inevitably, for this was now the standard procedure, patented.

The first real attempt at a working solution appeared when, in July 1962, invitations were received by a number of shipowners, Classification Societies and the technical press, to attend a demonstration of 'A low temperature integrated trial tank for the sea transport of Liquefied Natural Gas' in Oslo. The sponsors were Øivind Lorentzen, Shipowner, of Oslo and the Bennett Group Joint Venture, of Dallas; these two companies had, a year or so earlier, commissioned the Det norske Veritas research and development department to investigate LNG shipping techniques, primarily a spherical design.

Already, by 1961, DnV had requested they be allowed to develop an idea of one of their engineers, Bo Bengtsson, for a bi-axially flexible liner; the unique characteristic of this liner being that it embodied a method of folding the liner without stretching it, so that on contraction it could unfold elastically (see fig. 10n(vii)).

A trial tank of $32\,m^3$ had been built and tested with liquid nitrogen to demonstrate the design. In this tank the membrane, fabricated of 3 mm thick aluminium of very pure alloy, was screwed to an insulation system comprised of a grillage of timbers faced with plywood and filled with mineral wool. A layer of 80 mm polyurethane foam faced with polyester film provided the

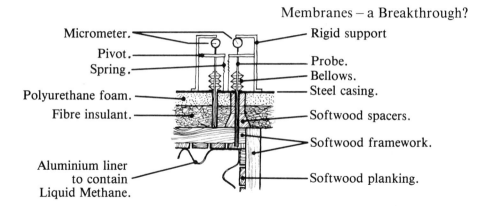

Fig. 11a. Section of tank with instruments to show relative movement.

secondary barrier protection for the mild steel outer casing which represented the inner bulk of the ship (fig. 11a).

Prior to the demonstration an extensive series of tests had been carried out on various liner shapes which included:

— measurement of bi-dimensional elastic properties;
— static and fatigue hydraulic tests
  (Pmax = 4 kg/cm²; Pdyn = 1.5 ± 0.4 kg/cm² × 100,000 cycles);
— simulated thermal fatigue testing
  (30 per cent extension in excess of thermal strain × 5,000 cycles – in one dimension).

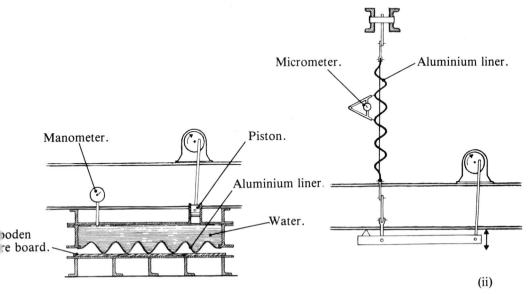

Fig. 11b. Set-up for pressure and extension fatigue tests.

123

The demonstration, to which Conch was understandably not invited, was obviously impressive; C.I. Kelly, correspondent for the *Petroleum Times,* and a keen observer of progress in this field of endeavour, wrote in the 10th August 1962 issue:

> 'The effect of shock-cooling was impressive. The visitors had seen inside the tank when warm on the first day – its elements, the absence of internal stays, the sturdiness of the geometric forms, the non-dependence of the liner on any pressure inside the tank to keep it against the planked surface. After lunch, liquid nitrogen from Norsk Hydro's synthetic ammonia plant was decanted at − 196° C from a tank-car into the test tank's liner which had not been pre-cooled. The behaviour of the liquid was seen through sight-glasses in the roof. Thermocouples placed at strategic points showed temperature changes throughout the structure; feelers [see fig. 11a] showed in micrometer gauges, set to zero in the warm condition, what movement had been stimulated at wooden spots and liner-surfaces. The latter moved about 0.3mm over a length of 4m from the warm position relative to the casing. Next morning, the cold-soaked assembly, about half full of liquid nitrogen, was pivoted along one short side and the other allowed to drop suddenly on to hydraulic jacks. These developed deceleration forces of about 1G on the whole unit. By special request this was repeated.
>
> 'The sponsors confidently expect LNG tankers built to their design and procedures to cost not more than 75 per cent of what is believed to be the cost of construction based on others' designs now accepted in this highly specialized branch of shipbuilding. . . . the c.i.f. price of this new commodity (LNG) in international trade should fall appreciably, should the sponsors' expectations materialize.'

Amongst the shipowners invited to the demonstration was René Boudet who was quick to see the potential of this new design; thus it was not long before the 'inventor' Bo Bengtsson found himself in the Paris offices of the newly formed 'Technigaz' company, a wholly owned technical subsidiary of Gazocean S.A.

## THE TECHNIGAZ SOLUTION[1/2]

Being convinced that the membrane concept was the LNG design of the future, and having acquired the design and patent rights from Lorentzen and the Bennett Group, Technigaz began an urgent development programme; this was based on the belief that (a) it was essential to substitute stainless steel for aluminium as a basic material of construction, because stainless steel

(i)    was a well proven cryogenic material;

(ii)   was readily available in commercial quantities;

(iii)  was easy to weld (it was already widely believed on reasonably good evidence, that aluminium, particularly of light scantling, was difficult);

(iv)  had a coefficient of expansion of about half that of aluminium which allowed a much wider spacing of corrugations;

(v)  had good corrosion resistance in a marine atmosphere; and

(vi)  had good fatigue resistance, which, indeed, improved at low temperatures;

and (b) the successful operation of a small prototype ship was the only way to provide convincing evidence that the design was both technically and commercially viable.

The main thrust of their work was now to develop a membrane design which could be reliably produced on a commercial basis without creating undesirable stress concentrations at the points where the corrugations crossed; and to design a similar detail for the tank corners. (1.0 mm was selected as a suitable thickness.)

A further fundamental design change was the incorporation into the system of a fully reliable secondary barrier, the tightness of which could be monitored in service; this took the form of a metallic barrier of identical form to the primary membrane.

After considerable experiment and testing the arrangement shown in fig. 11c was developed and installed in the 605 m³ methane/ethylene tanker *Pythagore*, which entered active service in May 1964. After a first trial loading, followed by an inspection by five Classification Societies, which included the USCG, she loaded a cargo at Nantes which was delivered to Canvey Island for partial cooldown tests on the new methane terminal.

An important factor in any membrane system is the atmospheric control of the spaces between the barriers and the inner hull–not only must the atmosphere itself exclude air and moisture but pressure differentials must be carefully maintained to avoid excess 'back pressure' which can force the membrane off its attachments to the insulation.

*Pythagore* was fitted with equipment to maintain the $N_2$ within these spaces at a dewpoint of less than $-60°C$; relief valves to prevent pressure *differentials* in excess of 30 gr/cm²; and relief to atmosphere at 100 gr/cm² (see fig. 11d).

After two LNG cargoes, *Pythagore* entered LPG/ethylene service and remained in this service until 1973–when she was converted to a refrigerated sardine carrier.

Only one real problem developed after just one or two trips and this was fatigue cracks which developed in the primary barrier at the bottom corners where there was clearly too much flexibility in the support system; local

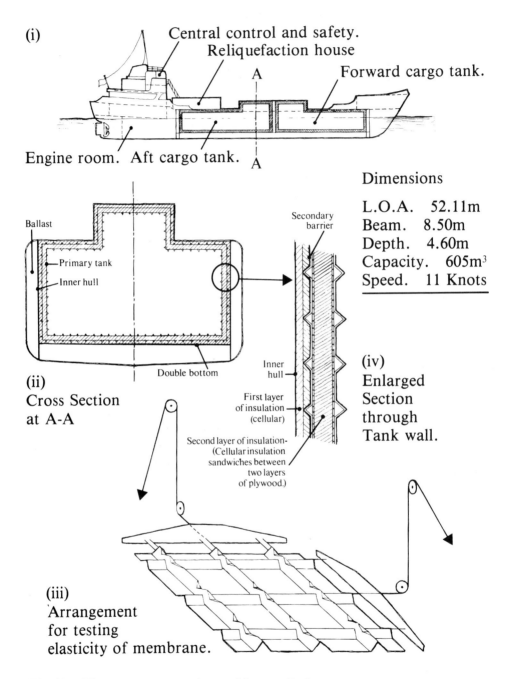

(i)

Central control and safety.
Reliquefaction house

A

Forward cargo tank.

Engine room.   Aft cargo tank.   A

Ballast

Primary tank
Inner hull

Secondary
barrier

Inner
hull

First layer
of insulation
(cellular)

Second layer of insulation-
(Cellular insulation
sandwiches between
two layers
of plywood.)

Double bottom

(ii)
Cross Section
at A-A

(iv)
Enlarged
Section
through
Tank wall.

(iii)
Arrangement
for testing
elasticity of membrane.

Dimensions

L.O.A.   52.11m
Beam.   8.50m
Depth.   4.60m
Capacity.   605m³
Speed.   11 Knots

Fig. 11c. First prototype membrane ship m.v. *Pythagore*.

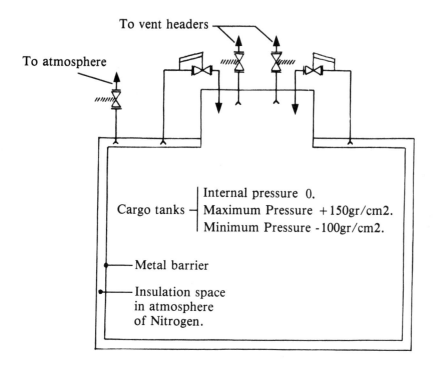

To vent headers

To atmosphere

Cargo tanks ⊣ Internal pressure 0.
Maximum Pressure +150gr/cm2.
Minimum Pressure -100gr/cm2.

— Metal barrier

— Insulation space
in atmosphere
of Nitrogen.

Fig. 11d. Protection of cargo tanks.

doublers were fitted and this cured the trouble, but the fault was significant in judging the suitability of the design for larger scale application.

Another single incident of interest was the incorrect opening, during drydock, of a valve connected to the insulation space which permitted this space to be 'pressurized' to well above its design condition and resulted in the detachment of a large area of membrane.

Flushed with the unqualified success of this venture, which caused enormous interest worldwide, M. Boudet anchored *Pythagore* for public inspection in Monte Carlo for the first LNG Conference, 'Petroleum and the Sea', in May 1965. But not before Conch had sued Gazocean for infringement of several of its patents and had 'entered' the offices of the builders, Duchesne et Bossiere, Le Havre, for the purpose of acquiring copies of all relevant drawings of the offending ship. The fun had started!

## THE CONCH SOLUTION[3]

Conch Methane Services had been set up in London after Shell had bought a

Plate 1. *Pythagore* at Monaco — 1965.

Plate 2. View inside cargo tank–*Pythagore*.

128

Plate 3. Detail of corrugation crossing–Technigaz membrane.

40 per cent stake in Constock (Chapter 6); its Research and Development department was headed by a Frenchman, 'Steve' Schlumberger, seconded from Société Maritime Shell in Paris and the very same individual who had introduced Gazocean to the advantages of carrying LPG in a part- or fully-cooled condition. Was it fate or a far-sighted management decision that led him to this position; for that matter, was it fate or the devil who allocated him an office overlooking Nelson's Column in Trafalgar Square? It will never be known, and the act was never forgiven!

Thus while Conch's Technical Director, Chuck Filstead, ex-Constock, assisted by the author, was concentrating on the vital and immediate job of seeing that the first LNG project went into operation in a smooth and trouble-free manner, and that the two ships associated with it performed likewise, Steve Schlumberger, assisted by Bob Jackson, also ex-Shell, were totally convinced that the future assuredly lay in the membrane and were authorized to allocate a 'substantial effort towards developing a viable design'.

Following much the same reasoning as Technigaz, Conch decided that stainless steel was the most suitable membrane material. Mainly because they gave high priority to the application of automatic rather than hand welding, and were also looking for ample margins in extensibility, Conch eventually rejected solutions based on dimpled or corrugated sheets and selected a combination of rotating trays, welded along their upturned edges and fixed to the insulation system at their centres by specially designed fittings to permit free rotation. The insulation system was, not surprisingly, the Conch balsa panel system – now a 'fully restrained' system rather than a series of individual panels attached to each other with expansion joints as on the *Methane Pioneer*; this was in itself a significant design change eminently suitable for a membrane application.

Prototype testing was, in retrospect, somewhat in advance of current thinking in Norway/France in that an attempt was made to simulate a ship's life cycle acceleration spectrum; this latter also required some original thinking on the part of the Classification Society (Lloyd's Register) who developed the following criteria:

'The response of a ship about 500 feet in length to the North Atlantic Sea state has been computed and, assuming the ship spends two-thirds of its life at sea, the following values are applicable to a ship life of twenty years:

| Sea State | No. of Cycles $\times 10^6$ | Bow Acceleration (g) |
|---|---|---|
| 0–3 | 15.28 | .03 |
| 4 & 5 | 28.26 | .195 |
| 6 & 7 | 14.64 | .375 |
| 8 & 9 | 3.24 | .600 |
| 10 & 11 | 0.34 | .756 |

'For fatigue testing purposes it is considered the above accelerations should be associated with a mean rolling angle of 15°. Assuming a tank 60 feet deep, 80 feet wide and a tank position of .4L forward of amidships, the following pressure cycle programme is obtained for a methane cargo:

| Pressure Range p.s.i. | No. of Cycles $\times 10^6$ |
|---|---|
| 3.5–21.0 | .34 |
| 5.0–19.0 | 3.24 |
| 7.0–16.5 | 14.64 |

'A test of $3.6 \times 10^6$ cycles with a pressure range of 3.0–21.0 p.s.i. is considered suitable.'

Having successfully met all the required laboratory scale tests, a prototype tank was fitted to the chartered collier *Findon*, the conversion being carried out at the Harland & Wolff shipyard, Belfast. The inside dimension of the

(i) Design for flexible corrugation intersection.

(ii) Design with dimple sheet.

REJECTED DESIGNS

FINAL CONCH DESIGN CONCEPT

Corrugations

Basic module

Support points

(iii) Final Conch membrane design.

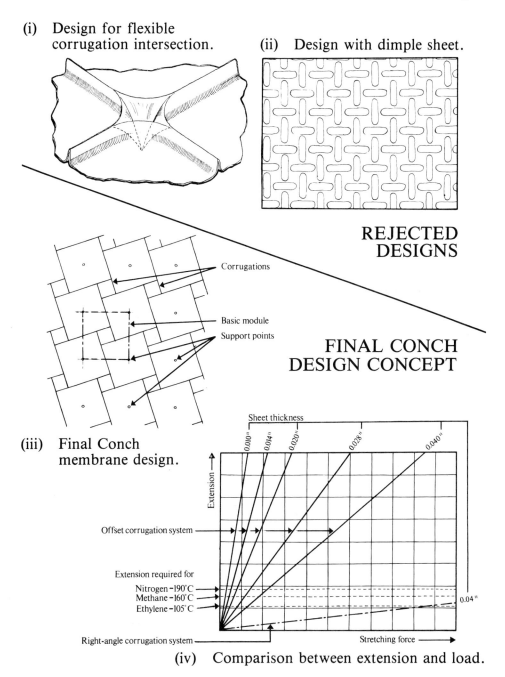

Sheet thickness

0.010"  0.014"  0.020"  0.028"  0.040"

Extension

Offset corrugation system

Extension required for

Nitrogen –190°C
Methane –160°C
Ethylene –105°C

0.04"

Right-angle corrugation system

Stretching force

(iv) Comparison between extension and load.

Fig. 11e. Conch Methodology.

mild steel tank, which was shaped to incorporate both fore and aft taper and bottom chamfer was 16 ft 6 ins long, 22 ft 0 ins/27 ft 6 ins wide and 24 ft 9 ins deep, with two 45°, 5 ft 6 ins deep bottom chamfers. After successful cooldown tests, *Findon* carried two cargoes of 125 tons of LNG to Canvey and three cargoes of ethylene to Arzew during the latter half of 1964.

It is, perhaps, amusing to note that visual examination of the membrane *after* the sea tests showed that the complete side of one tray had been un-welded to its neighbour and had not shown up as a leak!

Additional, more severe, fatigue tests carried out on pieces removed after the seagoing tests were completed, were not successful, however, and eventually the design was abandoned; but by this time further developments, both commercial and political, were afoot, which had a major impact on the fortunes of both Conch and Gazocean – and also on the development of membrane technology as a whole.

## THE CONCH OCEAN SOLUTION[4/5/6]*

By 1967, the patent dispute between Conch and Gazocean had reached the French courts, a preliminary judgement, lodgement of one million francs as surety, and was threatening to absorb a considerable amount of time and money; but by a piece of inspirational management, the affair was settled by the formation of a joint company, Conch Ocean (60 per cent Conch, 40 per cent Gazocean), registered in the Bahamas with a service company, Transgaz Services, in Paris. The objectives of this company were to develop the joint technology and expertise of both partners in the most efficacious manner. It was accepted at that time and without any reservation, that Conch had the best insulation/secondary barrier system, whereas Technigaz had the best membrane. The adaptation of one to the other should be relatively simple – and quick, which was an important factor. Conch Ocean would also be able to offer an independent tank design for the more conservative shipowners, and there were many, who preferred ships of more rugged design.

To add urgency to this 'joint venture', Esso had only recently rejected the TGZ *Pythagore* system on the grounds (justified) that it could not withstand the cyclic loads expected in a ship of 40,000 m³, of which they were about to order four for the Libya/Italy/Spain project; and Conch had failed in its attempts to participate in the Alaska/Japan project in competition with the Phillips Marathon offer, for which two membrane ships had recently been ordered – to be built at Kockums (of which more later in this chapter).

In a year of feverish activity Conch Ocean had successfully married their systems, having carried out prolonged fatigue tests on two boxes, each

* Now known as 'TGZ-Mk I' membrane, ref. Chapter 16.

containing a 'statistically acceptable' number of full size components, with a 1.2 mm thick membrane of improved design, manufactured on equipment suitable for large scale production and therefore fully representative of a commercially built vessel; additionally, the 'cold condition' had been simulated by prestretching the membrane. Having completed the test programme Conch Ocean received not only the required approvals from the Classification Societies – but also their praise for a job well thought out and executed.[5]/[6]

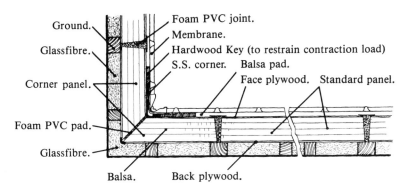

Fig. 11f. General arrangement of insulation system.

In early 1968, their results were presented to ATMA[4] and included a current forecast of expected cost savings (fig. 11g), now reduced to some 10–15 per cent.

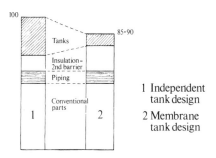

Fig. 11g. Cost comparison, Atma '68.

Before the year had ended, Gazocean ordered a 50,000 m³ vessel, the *Descartes*, to the Conch Ocean design, discussed in greater detail in a later chapter. But not much more than a year after that the partnership was dissolved for commercial reasons which were never fully understood by the technical participants to whom the dissolution was a great sadness. It was only a few weeks after the final break-up that Shell ordered five ships to the

Conch Ocean design for their Brunei/Japan project–*c'est la vie*!

One of the problems inherent in this design, now the TGZ membrane, was the cost of the balsa panel system. The TGZ Mk II sought to replace the balsa with PVC/ply sandwich panels and a simpler joint design, and undoubtedly this has reduced cost without sacrificing safety–the latter having always been the overriding consideration in the development of this system. A later design (Mk III) uses polyurethane foam blocks with a reinforced aluminium foil secondary barrier facing, developed in the USA by the General Electric Company.

But we are proceeding too fast–the date is still the mid-sixties!

## THE GAZ TRANSPORT SOLUTION[10/11]

Having decided that the cylindrical tank design used on *Jules Verne* was unlikely to develop into a commercially viable system, Gaz Transport concentrated on a membrane design based on 36 per cent nickel steel (Invar) which has a very low, almost indiscernible, coefficient of expansion. This virtually solves the temperature variation problems at a single stroke, leaving the hull deflections as the major concern, plus, of course, methods of attachment and the supporting insulation. Gaz Transport had also already decided that the secondary barrier would be of identical design to the primary–thus providing, as it were, a 100 per cent standby arrangement.

The insulation method adopted for the support of the membrane was a series of subdivided plywood boxes, factory-made on a production line basis to very exact tolerances; the boxes were filled with Perlite granular/powder insulation–inexpensive but effective. Purging of the air from the boxes, and circulating nitrogen within them and sampling for cargo leakage was made possible by arranging gauze covered holes in the sides of each box.

The maximum width of commercially available 0.5 mm Invar was about 0.4 m and this determined the width of both the boxes and the 'strakes', or strips, of membrane; length was no problem, so that it was possible for each strip to extend the complete length, or width, of the cargo holds. Each strip terminated in a 1.0 mm-thick corner angle piece which was 'keyed' into the ship's structure by a special arrangement of stainless steel tie-rods designed to minimize heat leak in those areas, thus providing for the necessary end restraint against membrane contraction (which was small*) and extension of the ship structure in bending.

* 5 kgf/mm² for *Polar Alaska*.

134

LNG spill on water 1957

LNG Pool — 1958 (See Ch.3)

**LAKE CHARLES TESTS**

MAPLIN SANDS SPILL TESTS 1980

pill Rate 3.2m³/ Sec.
oint of Ignition

Spill Rate 3.2m³/Sec.
Point of Ignition + 2.25 secs.

pill Rate 3.2m³/ Sec.
oint of Ignition + 9 secs.

Spill Rate 3.2m³/Sec.
Point of Ignition + 15 secs.

Note: 'Leisurely' ignition and burning characteristic;
white vapour cloud is water condensed from atmosphere

*(photos by kind permission of Shell Research Ltd)*

*Jules Verne* at Le Havre — 1965. (Since renamed "Cinderella".)

*Arctic Tokyo* loading in Alaska — 1969.

*Methane Progress* on trials — 1964.

*Methane Pioneer* at Lake Charles — 1959.

*Mostefa Ben Boulaid*–125,000m³, Technigaz design–built by La Ciotat for CNAN —1976.

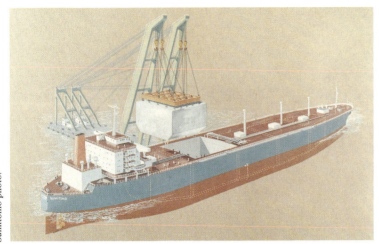

The Conch 125,000m³ design–as offered by Sumitomo shipyard.

25,000m³ sphere in transit to General Dynamics Shipyard.

*LNG Aquarius* on cold trials at Canvey Island Methane Terminal.

A wide range of laboratory tests investigated the system's fatigue life, Invar welding techniques and resistance to fatigue, the strength, compatability and thermal resistance of the insulation boxes, and leak testing and repair techniques. Once these had been completed successfully, prototype testing was carried out with LNG in a $25\,m^3$ test box. This test box was subsequently fitted to the forward deck of *Jules Verne* where it remained, filled with liquid nitrogen and subjected to normal ship accelerations, for about ten months.

Having passed all the required tests satisfactorily, and obtained the necessary Classification approvals, the first large scale commercial application was applied as a single layer of insulation boxes and one 0.5mm Invar membrane in the $30,000\,m^3$ LPG tanker *Hipolyte Worms*. This was completed by the CNIM shipyard in 1968.

This enterprise was only a qualified success, in that not only was difficulty experienced in maintaining adequate tightness of the membrane in service but, and this was possibly the main reason for this deficiency, considerable difficulty had been experienced in manoeuvring the long strips of Invar, with edges ready flanged in the workshops, into position in the ship.

Urgent attention was therefore given to improving installation techniques, which resulted in a method of flanging the strips (which were supplied in rolls) continuously in a special machine fitted in the ship's holds. The benefits of this breakthrough in production were immediate and could be seen in tangible form in the two $71,500\,m^3$ LNG ships already under construction at Kockums.

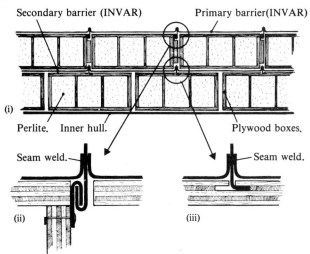

Fig. 11h.   Principles of Gaz Transport membrane.

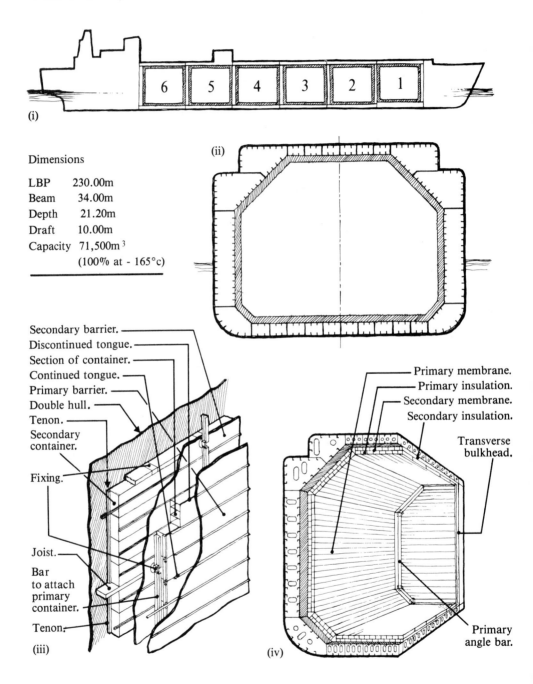

Dimensions

| | |
|---|---|
| LBP | 230.00m |
| Beam | 34.00m |
| Depth | 21.20m |
| Draft | 10.00m |
| Capacity | 71,500m$^3$ |
| | (100% at - 165°c) |

Fig. 11i. *Polar Alaska/Arctic Tokyo*

These two ships, *Polar Alaska* and *Arctic Tokyo*, entered service between Alaska and Japan in November 1969 and March 1970 respectively; they are still satisfactorily trading on that route (1992) – though not without having experienced two significant incidents in their early years. These incidents, which have been fully described in technical presentations[15] were briefly as follows:

(1) *Polar Alaska*: during the first return trip from Tokyo to Alaska – in ballast, with No. 1 tank filled 15–20 per cent with LNG to provide liquid for cooldown prior to loading, sloshing of the cargo caused part of the cargo pump electric cable support tray to break loose and become washed back and forth across the aft end of the tank. This perforated the membrane in several places causing liquid leakage into the secondary barrier space – which remained intact. Repairs took less than one week.

Prior to the above repairs an excess pressure of about $100\,gm/cm^2$ was accidentally applied to the space behind the primary membrane causing considerable deformation of the membrane and requiring complete scaffolding of the tank for repairs. (Time for repair was approximately six weeks).

(2) *Arctic Tokyo*: although carrying a 'heel' in the tanks was discontinued after the *Polar Alaska* incident, the practice was resumed in March 1971 after tests had indicated that this could safely be done. However, the vessel hit bad weather in ballast later that year and again leakage was detected in the No 1 tank insulation space.

It was subsequently found that local deformation of the membrane and supporting insulation boxes had occurred at the aft tank corners at the liquid surface (20 per cent full) level. Leakage occurred at one of these points. A maximum 5 per cent fill level was then imposed and no further problems of this kind have been experienced on these two ships.

When applied to ships of 125,000m$^3$ capacity, this design developed problems due to fatigue at the points where the primary barrier tie rods passed through the secondary barrier at the tank corners. This was particularly pronounced at the lower and upper areas where sloshing forces are greatest and ship bending deflections highest. Modifications to the design were made to existing ships to provide a more flexible connection but a new design has now been developed which eliminates point penetrations (see Chapter 16).

*Polar Alaska* and *Arctic Tokyo* are due to be replaced by two 85,000m$^3$ ships of the IHI/SPB design in 1993/94 (see Chapter 16).

Natural Gas by Sea

Natural Gas by Sea

Natural Gas by Sea

[1] 'Le navire méthanier-éthylénier *Pythagore*' by J. Alleaume. Association Technique Maritime et Aéronautique, 1964.
[2] 'Le transport maritime du GNL et de l'éthylène–la technique *Pythagore*' by J. Alleaume. Petroleum and the Sea Conference, Sect III No 311, 1965.
[3] 'Sea transportation of LNG–future trends' by R. G. Jackson and R. C. Ffooks. Petroleum and the Sea Conference, Sect III No 302, 1965.
[4] 'La conception et la mise au point d'une technique du cuve membrane pour le transport du gaz naturel liquéfié' by G. Massac and R. C. Ffooks. Association Technique Maritime et Aéronautique, 1968.
[5] 'Testing and technology of models of integrated tanks for LNG carriers' by R. G. Jackson and M. Kotcharian. International LNG Conference, Chicago, 1968.
[6] 'Evaluation of the fatigue strength of integrated tanks for LNG ships' by D. J. Burns, R. G. Jackson and J. G. Kalbfleisch. International Conference on LNG, organized by the International Institute of Refrigeration and the British Cryogenics Council. London, March 1969.
[7] 'LNG tanker technology according to the Technigaz designs' by J. Alleaume and F. Alvarez de Toledo. LNG/LPG Conference, London, 1972.
[8] 'Development of the Technigaz integrated tank system for methane carriers' by J. M. Chauvin. Gastech 75, Paris, 1975.
[9] 'The General Electric-Technigaz Mark III containment system' by J. Roni and J. M. Chauvin. Gastech 78, Monte Carlo, 1978.
[10] 'Concept of model testing of Gaz Transport/Gaz de France membrane tank for methane tankers' by M. Guilhem and L. L. Richard. International LNG Conference, Chicago, 1968.
[11] 'Larger membrane tankers' by A. Gilles and J. Guilhem. LNG/LPG Conference, London, 1972.
[12] 'Development of the 3-D containment systems' by Dr J. L. Waisman and A. Gilles. Gastech 76, New York, 1976.
[13] 'Adapting a new containment system to a shipbuilder's product' by F. P. Eisenbiegler and J. D. Mazzei. Gastech 76, New York, 1976.
[14] 'A new Invar membrane containment system for construction of LNG carriers' by A. Gilles and Dr J. L. Waisman. LNG 5, Dusseldorf, 1977.
[15] 'Alaska to Japan LNG project — Kenai revisited by J. Horn, Paul Tucker and W.B. Emery II — LNG 4, Algiers 1974.

138

# 12
# Two More Projects and a Speculative Ship

We must now go back a few years to the end of 1965; Conch, flushed with the success of the start up of the Algeria-UK project, are preparing for further LNG supplies to the UK from Nigeria. Tenders are out for two ships of twice the capacity of *Methane Princess* and *Methane Progress* and a company has been set up in Lagos.

At the time intense activity centred around the impending contract between the Union-Marathon, later Phillips-Marathon Group and Tokyo Gas to ship about 140 million cubic feet of gas per day into Japan. Conch were bidding for a principal role in the liquefaction and shipping for this Japanese importation project.

Esso were in the process of developing their Libyan gas reserves with potential markets in both Italy and Spain – so why should Conch, who, after all, now held all the aces in this game, not participate as principals in this project too?

Other LNG projects were also being pursued.

In addition, the patent litigation against Gazocean, who now had not inconsiderable ambitions in this field, appeared to be moving in Conch's favour.

It was in October 1965 when the first 'disaster' struck with British Petroleum's discovery of gas in the British Sector of the North Sea, only fifty miles off the Yorkshire coast. The Nigeria-UK project collapsed forthwith. However, the results of the shipyard bids were interesting in that after receiving what appeared to be disproportionately high estimates from the UK yards, bids were sought from several European yards with similar results— one German shipyard, having offered the best price, hastily withdrew it when interest was shown! Subsequently proposals were confidently sought from three Japanese yards currently quoting tanker prices over 20 per cent below European – but their bids for LNG ships were higher than the UK/European prices.

Nor were Conch very much more fortunate in achieving participation in either the Esso or Alaska projects – as will be seen.

## THE ESSO PROJECT[1]

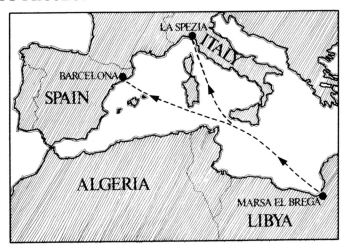

Fig. 12a. Esso Libya-Italy-Spain Project.

Esso signed their agreements with Libya, Italy and Spain to deliver 235 MMSCF and 110 MMSCF per day respectively in the final weeks of 1965. This project, due to start in 1968, required four ships of 250,000 bbls (± 40,000 m³) capacity and 18 knots service speed. The size and speed was determined by considerations of 'supply logistics, customer commitments and economics'; their specific design was chosen '. . . with major emphasis on performance and reliability'.

After careful examination of the available LNG containment systems, their relative costs, licensing conditions and potential reliability in service, ˙sso decided to develop their own independent tank design and, adopting a

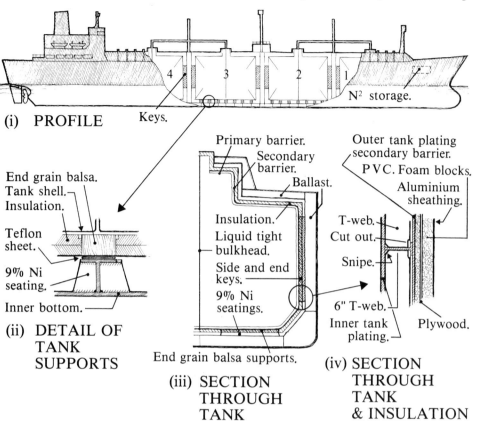

(i) PROFILE    Keys.

N² storage.

End grain balsa.
Tank shell.
Insulation.

Teflon sheet.

9% Ni seating.

Inner bottom.

(ii) DETAIL OF
TANK
SUPPORTS

Primary barrier.
Secondary barrier.
Ballast.
Insulation.
Liquid tight bulkhead.
Side and end keys.
9% Ni seatings.

End grain balsa supports.

(iii) SECTION
THROUGH
TANK

Outer tank plating
secondary barrier.
P V C. Foam blocks.
Aluminium sheathing.

T-web.
Cut out.
Snipe.

6" T-web.
Inner tank plating.
Plywood.

(iv) SECTION
THROUGH
TANK
& INSULATION

very conservative approach, opted for a double walled, divided aluminium tank. Largely to avoid the Conch patent portfolio, they adopted a vertical side key arrangement in which, by spreading the loads over a large area, they could use timber as the load bearing insulation. The insulation, comprising an assemblage of PVC blocks in two layers with glass fibre fitted between adjacent blocks to accommodate the expansion and contraction of the tanks, was attached to the outer tank shell; this location for the insulation not only avoided an important feature of the Conch basic patent, but has the advantage that any weak points in the insulation were much less likely to show up as cold spots in the inner hull, besides which it was possible to examine the outer surfaces with the tanks in the cold condition.

The order for the four ships was allocated to the Italcantieri shipyard in Genoa (3) and the Astano yard in El Ferrol, near La Coruña in North West Spain (1); Italcantieri were nominated as the lead yard, where the Esso design development and supervisory team, which included a Conch representative as adviser, was based. The responsibility for the final design details of the

141

cargo tanks and subsequently their construction–and insulation–was allocated to Chicago Bridge & Iron. The building of these tanks presented considerable difficulties, mainly due to the awkwardness of the double walled construction, but not helped by the fact that the forward tank was designed with a double curvature in the side walls to obtain maximum volumetric efficiency; the main problem lay in establishing acceptable welding sequences for the more complicated areas–and this particularly applied to the heavy extruded I-section which ran along the bottom centreline.

The internal tank structure comprised standard tanker horizontal stringers and deep vertical web stiffening–unlike the Conch tanks which sought to eliminate all vertical girders with the intention of reducing the stresses induced by vertical temperature gradients. Thus the Esso tanks needed careful control of temperature gradients at all times which was achieved by distributing a series of spray cooling nozzles throughout the tanks.

All four ships were fitted with Carter submerged pumps, two in each tank having a combined discharge rate of 4,000m³/hour; furthermore each tank was fitted with a valve in the centreline bulkhead, to provide a standby facility in the event of pump failure and, additionally, with an eductor which could be used in conjunction with pumpage from an adjacent tank.

The principal particulars of the ships were as follows:

| | |
|---|---|
| Length (BP) | 109.97 m |
| Beam | 29.26 m |
| Depth (deck at side) | 18.47 m |
| Depth (trunk deck) | 21.92 m |
| Draft | 8.63 m |
| Cargo capacity at 98.5 per cent | 40,000 m³ |
| SHP | 15,000 |
| Speed | 18 knots |
| No. of tanks | 4 |

The first ship, *Esso Brega*, underwent trials without incident and entered service in October 1969; the second, *Esso Portovenere* in March 1970. *Laieta* and *Esso Liguria* followed later the same year.

The project start-up was beset with unforeseen problems; firstly two fires in the pipeline supply system, followed by local political unrest, so that the ships were not fully employed until well into 1971, after which date they operated without any technical problems, although the two projects did experience hold-ups for political reasons.

Thus, while the Esso design can be deemed to have been entirely successful albeit extremely conservative, it was never likely to be repeated due to its high initial capital cost and complex construction. The four ships are reported to have cost $104 million.

## THE ALASKA-JAPAN PROJECT[2-8]

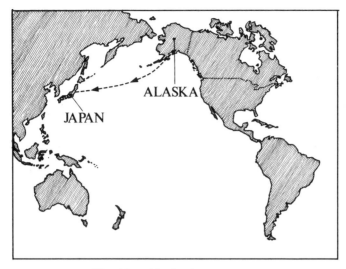

Fig. 12c. Alaska-Japan project.

This, the first Japanese import project, was won in a straight commercial contest between Conch and Phillips-Marathon, the latter undoubtedly deriving significant financial benefit from, but accepting substantial risks in, the acceptance of a completely new and very largely untried LNG ship design – the Gaz Transport membrane (see page 134)—currently quoted at an extremely competitive price by the Kockums shipyard in Malmö, Sweden. A figure of $25 million for each of the two ships was reported at the time. It was further reported that the shipyard lost on the first ship but recouped these losses on the second!

Brief particulars of the two ships are as follows:

| | |
|---|---|
| Length (BP) | 230.00 m |
| Beam (ext.) | 33.99 m |
| Depth | 21.20 m |
| Draft | 9.50 m |
| Cargo capacity | 71,500 m³ |
| SHP (service) | 18,000 |
| Speed (service) | 17 knots |
| No. of tanks | 6 |

The ships were sized conservatively for this rather severe service, having an ample speed margin based on a 345-day operating year; they were provided with ice strengthening forward and designed to meet all current requirements of the American Bureau of Shipping and the United States Coast Guard;

side thrust units were fitted at the bows to assist in docking at Kenai where tugs were not expected to be regularly available.

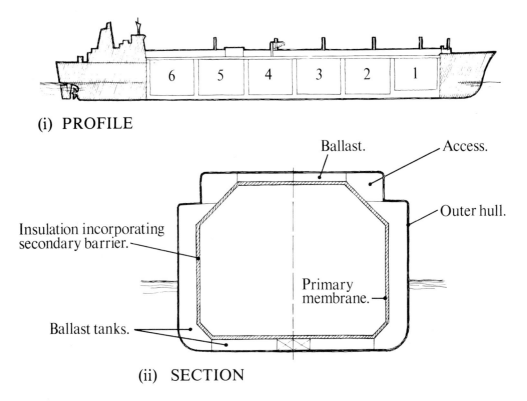

(i) PROFILE

(ii) SECTION

Fig. 12d. Arrangement of *Polar Alaska* and *Arctic Tokyo*.

With the benefit of the experience of *Hypolite Worms*, considerable efforts were made towards improving both welding and fabrication of the membrane –much of this work being carried out in the shipyard laboratories, with the result that the standard of construction was entirely satisfactory to all concerned.

After a prolonged 'maiden' voyage via Cape Horn to Alaska and incident free cooldown which took eleven days, *Polar Alaska* entered LNG service on 15th October 1969. Perhaps the first loading was not entirely incident free as at one stage one of the ballast crossover valves jammed and the ship took up an alarming list at the jetty. Some of the less phlegmatic crew members are reported to have hurriedly left the ship–by the gangway, as the ice floes did not exactly encourage diving into the water. The *Arctic Tokyo* entered service on 1st March 1970 after a cooldown/test programme lasting only six days.

The only significant problems which these two ships appear to have met in their twenty-five years contract service are (i) sloshing damage in their early years[8], (ii) local cracking of the deck around the tank openings and, later (iii) some inner hull leakage.

It was the sloshing damage which caused considerable concern at the time. Quoting from reference 5, this particular incident manifested itself

> '... during the second loading in Alaska of *Polar Alaska* ... with all tanks on average 80 per cent filled, it was observed, 19th November, 1969, that
>> (a) a gas alarm registered in the insulation space of tank No. 1;
>> (b) an abnormal drop in the temperature of the secondary barrier, at the aft bulkhead, to that of LNG ($-160°C$);
>> (c) an increase in pressure in the insulation space.
>
> These facts could only mean a loss of liquid tightness in the primary barrier of tank No. 1; loading was stopped immediately and the cargo from No. 1 tank distributed into the remaining tanks which had sufficient ullage.'

Subsequent inspection revealed the damage already mentioned in Chapter 11.

The end result, after much research into the sloshing phenomenon, was to impose strict limitations on the free surfaces allowed in membrane tanks; this in itself did not completely solve the problem in the case of the 125,000m³ class ships which entered service some ten years later, although more recent design changes appear to have done so (see Chapter 16).

The fact that the two ships have operated regularly on this arduous run for the whole of their twenty-five years contract is testimony enough to the fundamental integrity of this design.

## THE FIRST SPECULATIVE SHIP[9]

If the reader has not deduced from the last chapters that the battle between the containment systems had been joined then he must be made aware of this fact now. By the late 1960s there were two membrane designs and one viable independent tank system – the pioneer of Natural Gas by Sea – each convinced that their system was the superior, each trying to convey this message to potential shipowners, each attempting to recoup their development costs from the royalties which would accrue from ship orders – even if they were unable, or unwilling, to participate as principals in any of the projects which were now appearing in ever increasing numbers.

Never one to hang back, René Boudet, Managing Director of Gazocean, Paris, now a rapidly expanding LPG shipowner, and with a shotgun marriage on his hands in the form of Conch Ocean – but, by the middle of 1968, with a new and approved membrane design in his pocket, indicated his confidence in the design by placing an order with Chantiers de l'Atlantique for a 50,000 m³ LNG ship to the Conch Ocean design.

The original concept behind the decision to order this ship, *Descartes,* was ingenious; some said, with the benefit of hindsight, ingenuous; but it was surely courageous.

The trade was undoubtedly expanding, the technology was 'proven', estimates at the time pointed clearly to the need for at least eighty LNG ships by 1980 to satisfy the potential thirst of the USA and Japan combined – with Europe's needs not far behind.

Every graph seemed to be ascending at an angle of 45°! New prospects and projects were developing weekly – or so it seemed. In this perhaps over-rarified atmosphere there seemed to be real potential for the entrepreneur with a vessel which could fill the inevitable gaps caused by, for example:

(a) late delivery of ships dedicated to a specific project;
(b) the inevitable breakdown of ships already trading in a highly geared project;
(c) unforeseen seasonal or other more permanent, but unpredicted, expansions of projects;

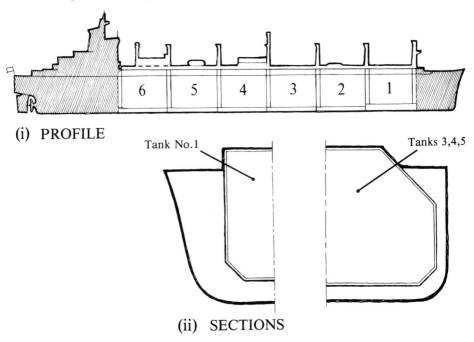

(i)  PROFILE

(ii)  SECTIONS

Fig. 12e.  *Descartes* – 50,000 m³ LNG capacity.

for it was well enough appreciated that most LNG projects, capital intensive as they were–and on a grand scale, had very little in-built margins either physically or contractually, for the unforeseen non-performance of any one unit.

*Descartes* was therefore designed and built to slip into any of the existing, and immediately foreseeable, LNG schemes at short notice, and this included the huge Shell-Brunei/Japan project which was at that time firming up quite rapidly. To provide additional flexibility she was designed for cargoes of density up to 0.65 and provided with a reliquefaction plant to enable her to engage in the LPG trade. The possibility was also foreseen that she might be 'shared' by all LNG project 'owners' as a tangible insurance policy.

In the event no speculative opportunities arose for *Descartes* but, by the time she was ready for service the Algerian/Distrigas, Boston project had matured – with the assistance of Gazocean – and she was allocated to this service, until 1979 after which she has been employed on deliveries between Algeria and France. By the end of 1991 she had completed over 448 voyages – not bad for a speculation!

In addition to their intensive efforts to develop and market their membrane design in association with Conch, Technigaz had been quietly developing a spherical design in competition with Kvaerner Moss; details of the prototype, *Euclides*, which entered service in 1971, are discussed at greater length in Chapter 13.

Thus by the end of 1971 there were five containment systems in service, and ships' size had reached 71,500 m³ capacity–and the Conch Ocean partnership had been dissolved.

[1] 'LNG from Libya' by A. Delli Paoli, F. S. Pramuk and R. W. Sage. LNG 2, Paris, 1970.
[2] 'The Phillips-Marathon Alaska to Japan LNG project' by W. L. Culbertson and J. Horn. International LNG Conference, Chicago, 1968.
[3] Alaska to Japan LNG project start-up and operational history' by LeRoy Culbertson and W. B. Emery II. LNG 2, Paris, 1970.

147

[4] 'Enseignments tirés de la construction et de la mise en service des méthaniers *Polar Alaska* et *Arctic Tokyo*' by J. Guilhem and L. Richard. LNG 2, Paris, 1970.

[5] 'Construction et mise en service des méthaniers *Polar Alaska* et *Arctic Tokyo*' by J. Guilhem and P. Jean. Association Technique Maritime et Aéronautique, 1970.

[6] 'Operating experience with LNG tankers' by R. J. Wheeler and W. B. Emery II. API Conference, 1971.

[7] 'Alaska to Japan LNG project – Kenai revisited' by J. Horn, Paul Tucker and W. B. Emery II. LNG 4, Algiers, 1974.

[8] 'LNG ship operations Northwest Pacific from October 1969 to the present' by L. R. Jamison and R. D. Yuill. Gastech 75, Paris, 1975.

[9] 'Operating experience on the methane carrier *Descartes* as from the first trip' by A. Ph. Détrie. LNG 3, Washington, 1972.

# 13
# The Spheres

The sphere was one of the favoured shapes for containers in the early conceptual design days: it was simple, notch free, and had been successfully used for the storage of gases ashore, under pressure, for many years. It was reliable and could be designed and built economically according to existing and well tried codes.

Naval architects, however, did not much like spheres–they utilized the enclosed hull space very inefficiently and, if they were allowed to project through the deck, destroyed the integrity of one of his most treasured possessions. Furthermore, it is all very well supporting a sphere on legs ashore – on solid concrete foundations – but the flexible structure of a ship and its tendency to roll and pitch at substantial angles, posed seemingly insuperable problems, which appeared to outweigh by far any advantages available from simplicity of analysis and economy of construction of the sphere itself.

Despite this inherent resistance from the traditionalists–who in reality have not played a very significant part in the development of LNG technology –the Norwegian Kvaerner Group, looking for a fundamentally new approach in the mid-1960s, believed that there was considerable potential in a spherical system. Their conviction was based on satisfactory experience with LPG ships of this design — already built by one of the shipyards in their Group and proving to be very successful in service, and also on the very significant

Natural Gas by Sea

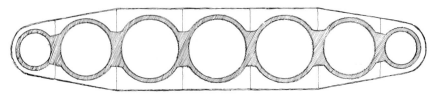

(i)   The First Patent (Hans Lorentzen - 1955)

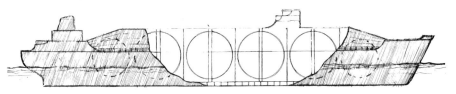

(ii)   The First Spherical Ship proposal - circa 1956

Fig. 13a.

support which they were able to obtain from what was then the Maritime Advisory Service of Det norske Veritas; this latter group were able to provide the vital in-depth structural design and analysis capability which, as will be seen, was essential to the successful development of the spherical containment system.

After several years of intensive design and development work, 'The design of an 88,000 m³ LNG carrier with spherical cargo tanks and no secondary barrier' was introduced at the LNG 2 Conference in Paris in October 1970.'[1]

'This paper describes the initial design of an LNG-tanker with special attention to the problem of achieving an increased safety of the primary tank so that the secondary barrier can be excluded. The five spherical cargo tanks are supported at the equator by a continuous cylindrical skirt. The cargo tank system is an integral part of the hull, and an extensive stress analysis is carried out to obtain the response of the cargo tanks. The most advanced methods have been applied to evaluate wave loads as well as stresses in the structures. The cargo tank material, 9 per cent Ni-steel, has been subjected to extensive tests to clarify the notch toughness properties for the thicknesses involved, as well as the fatigue data. The strength of the ship permits the carriage of ethylene, LNG and ammonia as alternative cargoes. Aluminium as an alternative cargo tank material is being investigated and preliminary results indicate that this material may offer an attractive alternative.'

150

This introduced a completely new element into the LNG scene and, with it, not a little scepticism. Orders were already placed, so the paper said, for three ships and one option–but where was the prototype? In fact, no prototype in the form of *Methane Pioneer*, *Beauvais* or *Pythagore* was contemplated. The Kvaerner/Moss Group felt sufficiently confident in their design to go straight for the commercial size ship–or ships. This was a bold step.

Existing regulations allowed the secondary barrier requirement to be omitted for pressure vessels – but was a sphere of this size (over 30m. diam.) and scantling, designed to operate at, or close to, atmospheric pressure and integrally connected to the hull, really a true pressure vessel? Was the standard of structural analysis and science of fracture mechanics sufficiently well developed to be certain that cracks in the sphere would be detected long enough in advance for leakage to be contained by the 'drip tray', or 'small leak protection', provided? Surely the equatorial ring support could not be relied upon completely when one of the principal welds could only be made, and inspected, from one side? And what about the torsional strength of the ship with such a large percentage of the effective deck area missing? Many such questions and more sprang to the minds of the audience at this conference.

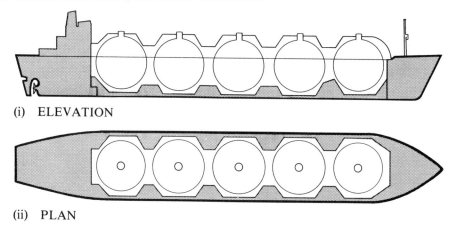

(i)  ELEVATION

(ii)  PLAN

Fig. 13b. Kvaener-Moss design for 88,000 m³ LNG ship.

Kvaerner Moss stood their ground, however–their studies, and supporting test work had been thorough. As described in their paper they had examined in exhaustive detail:

(1) *Wave loads and ship motions*: accelerations, shear forces, bending moments and torsional moments, other wave loads, i.e. dynamic bottom pressures and green sea effects, on the sphere covers. Liquid motions in the tanks.

(2) *Tank shell and supports*: the connection between the tank and support, or equatorial ring, being a vital component, had been subjected to both finite element and photo elastic analysis, and fatigue testing at full scale.

(3) *Structural analysis* had taken into account thermal stresses in the equatorial zone; thermal stress in the tank shell; torsional response of the hull and its effect on the spheres and supports; stability of the tank and supporting skirt under all foreseen loadings.

(4) *The arrangement* of the ship itself, and its insulation and cargo handling system were all carefully considered.

(5) *9 per cent Ni steel* had been investigated in detail as to its fatigue properties – and most particularly, its crack propagation characteristics under a range of stress levels, i.e. a fracture mechanics study in depth, this being the key to the whole design concept.

Kvaerner Moss summarized their presentation as follows:

'The most important tank conceptions for LNG tankers are:
  (i)    Membrane tanks
  (ii)   Self-supporting prismatic tanks
  (iii)  Horizontal cylindrical tanks
  (iv)   Vertical cylindrical tanks
  (v)    Spherical tanks.

'According to the Rules of Det norske Veritas LNG tankers equipped with pressure vessels of type (iii), (iv) or (v) may be built without a secondary barrier.

'The main reasons for choosing spherical tanks for the present design are fourfold:
  (i)    The risk for a collision to cause puncture of a cargo tank is minimized.
  (ii)   The entire tank structure lends itself comparatively well to theoretical calculation of stresses and carries loads mainly by membrane action.
  (iii)  The secondary barrier may be omitted.
  (iv)   The design is economical as regards steel weight and manufacturing costs in relation to cargo capacity.

'In addition to these main points there are several advantages associated with the use of spherical tanks. These include the large reserve strength for increase of the internal pressure which may be utilized for pressure boosted emergency discharge of the cargo. Also in other cases of emergency such as fire, the ability of resisting elevated pressure is most valuable. The tanks can also sustain a reasonable underpressure, representing a safety margin in case of faulty operation during e.g. discharge.

'The strength of the cargo tanks can easily be increased, merely by increasing the plate thicknesses, whereby heavy cargoes may be transported. For this particular project, where buckling properties control the overall shell thickness of the cargo tanks, one has automatically achieved tanks which are strong enough to carry LPG and ammonia.

'Among the difficulties experienced in the design are those related to stability of the ship. As is evident . . . the centre of gravity is relatively high above the keel and this calls for special design measures.'

The method adopted for supporting the spheres was a continuous cylindrical, stiffened, skirt attached to the equator by a special extrusion (fig. 13c) such that the sphere could freely expand and contract with minimum

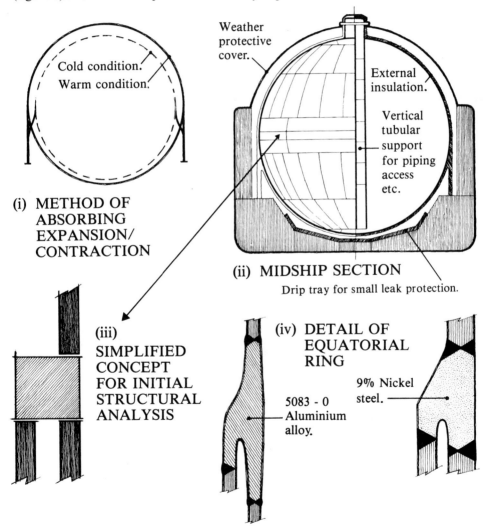

(i) METHOD OF ABSORBING EXPANSION/ CONTRACTION

Cold condition.
Warm condition.

Weather protective cover.

External insulation.

Vertical tubular support for piping access etc.

(ii) MIDSHIP SECTION

Drip tray for small leak protection.

(iii) SIMPLIFIED CONCEPT FOR INITIAL STRUCTURAL ANALYSIS

(iv) DETAIL OF EQUATORIAL RING

9% Nickel steel.

5083 - 0 Aluminium alloy.

Fig. 13c. Kvaerner-Moss sphere design and support concept.

loads being applied to the shell. The skirt is, in turn, welded integrally to the hull supporting structure; it is thus subjected to the ship's deflections and is designed to absorb them.

It can be seen from the illustration (fig. 13c[i]) how the relatively flexible skirt can accommodate horizontal shrinkage of the spheres without introducing unacceptable bending loads into the shell of the sphere–an important characteristic of any spherical support design.

One of the characteristics of LNG containment system designers is that faced with serious competition, they rise to the occasion–and quickly.

If shipowners were now ordering spherical designs straight from the drawing board–then spheres they would be offered.

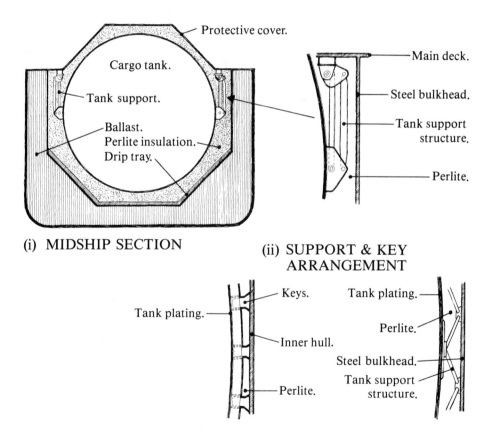

Fig. 13d. Technigaz sphere and support details.

M. Boudet of Gazocean and Technigaz was once again first in the field with a prototype, the 4,000 m³ *Euclides* (1971) which could, after completing its evaluation tests on LNG, be used in his diverse and growing gas carrier fleet for either LPG or ethylene cargoes. The Technigaz sphere[2] differed from the Moss only in its support system – indeed, it is only in the support system that any one sphere can vary from another. In an attempt to dissociate the ship's structural deflections from the sphere and, at the same time, accommodate relative expansion and contraction, the French design adopted a somewhat complex form of linkages which involved slinging the sphere from the deck (see fig. 13d).

This arrangement proved to be reasonably successful in the small prototype but, in developing designs for ships larger than about 50,000 m³ it became clear that the general complexity and cost would become too great for it to be commercially attractive; furthermore, the discrete support points would present technical problems. The design therefore became restricted to ships below 50,000 m³ in capacity and this effectively ruled it out for any major LNG project. It has now disappeared from the scene.

The next design to appear was that of the Chicago, Bridge & Iron Company, USA who capitalized on their vast experience of shore based spheres. Their system, fig. 13e, envisaged a large number of individual legs attached to the sphere in a conventional manner, but with anti-roll and pitch keys at the equator. The design did not find much favour even though it was subsequently modified, and improved, in conjunction with Newport News shipyard, to incorporate a double ring girder from which the tank was supported and keyed against horizontal rolling and pitching forces.

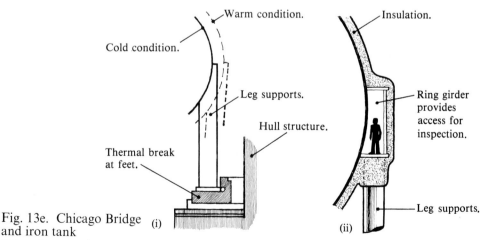

Fig. 13e. Chicago Bridge and iron tank support concept.

Another development from the USA at this time was a design produced jointly by Pittsburgh des Moines (USA) and Gaz Transport (France), drawing heavily from the latter's experience with *Jules Verne*. It was not a true sphere, but is mentioned in this section because it was stimulated by the spherical designs being developed at that time and also the current predisposition by United States shipyards towards independent, rather than membrane, designs.

The PDM/Gaz Transport design[3] was part sphere, part cylindrical, part cone and, as such, had advantages over the sphere in volumetric efficiency and ship stability but its weakness lay in designing suitable locating/keying

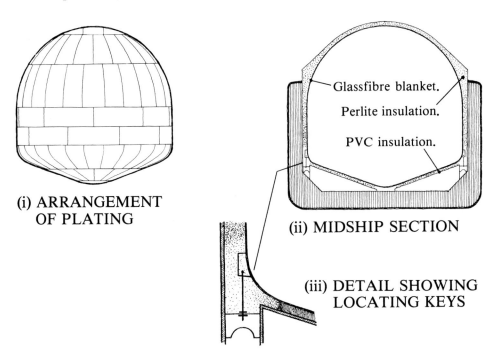

(i) ARRANGEMENT OF PLATING

Glassfibre blanket.

Perlite insulation.

PVC insulation.

(ii) MIDSHIP SECTION

(iii) DETAIL SHOWING LOCATING KEYS

Fig. 13f. Pittsburgh des Moines/Gas Transport 'Spherical' design.

arrangements, around the lower corners and the difficulty in providing effective insulation and secondary barrier protection at this location. Whether it was this reason alone, or the more general misfortunes that later befell PDM in the LNG containment field, which led to its abandonment is not known; at any rate the design is now firmly relegated to the list of 'also rans'.

At the same 1973 LNG conference the Spanish design group, Sener, presented their own version of a true spherical design.[4] This incorporated what appeared to be a somewhat complex supporting skirt, conceived in such a manner that 'it significantly reduces and practically eliminates' the bending moments induced in the tank shell at the equatorial ring due to contraction of the sphere when cooled. This is achieved, as can be seen from the illustration (fig. 13g), by arranging a cantilevered connection at the top of a double-wall skirt.

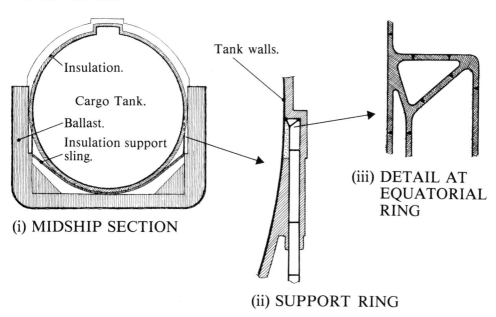

Insulation.

Cargo Tank.

Ballast.

Insulation support sling.

Tank walls.

(iii) DETAIL AT EQUATORIAL RING

(i) MIDSHIP SECTION

(ii) SUPPORT RING

Fig. 13g. Sener Sphere.

Apart from this feature the basic design concept of leak before failure, based on the principles of fracture mechanics and a detailed three-dimensional finite element analysis, was very similar to the Moss design. It was this similarity which so incensed the Moss Group, and resulted in prolonged patent litigation.

Confidence in the Sener system was such that a new shipyard, Crinavis, with a capacity for ships up to 300,000m³, was designed and part constructed in Algeciras, Spain for the express purpose of building this single design.

A 5,000³ prototype *Sant Jordi* was built in 1977 (at the Astilleros Tomas Ruiz de Velasco shipyard, Bilbao), with 9 per cent Ni steel tanks, and successfully tested with LNG, thus proving the technical capability of the system. But sadly no ship was ever built in the new Crinavis shipyard which in the event was never completed.

In 1982 yet another spherical tank design emerged (fig. 13h), this time from Japan[9]. The support system was developed by the Hitachi Zosen shipyard in conjunction with Chicago Bridge. The influence of a practical shipbuilder in their adaptation of the original CB&I design (fig. 13e) is clearly evident.

Making use of the double ring girder arrangement at the equator, the HZ/CBI sphere is supported on 32 stumps instead of long vertical legs and arrives thereby at what seems to be a compact, stable and practical layout, suitable for ships of 125,000m³ and above. None have been ordered however.

A small prototype of 1,100m³ capacity, incorporating one aluminium alloy spherical tank and one 9 per cent nickel prismatic tank was built by Hitachi and tested in 1974. The ship, *Sankyo Ethylene Maru,* was fully instrumented with strain gauges, accelerometers, etc, and the results made public in the Journal of the Society of Naval Architects of Japan (an English language summary being available as part of the shipyard brochure series).

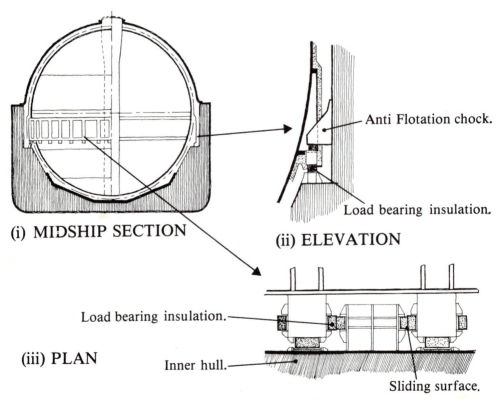

(i) MIDSHIP SECTION

(ii) ELEVATION

Anti Flotation chock.

Load bearing insulation.

Load bearing insulation.

(iii) PLAN

Inner hull.

Sliding surface.

Fig. 13h.  Hitachi Zosen/Chicago Bridge and Iron Sphere.

The real value of a prototype has always been open to question, but it will be noted that almost every design group has felt obliged to build a seagoing prototype of some kind before feeling sufficiently confident to embark on a full scale commercial venture. Kvaerner Moss and Gaz Transport (Membrane) are two notable exceptions – and, ironically, the two most successful designs today. Perhaps there is a moral here somewhere.

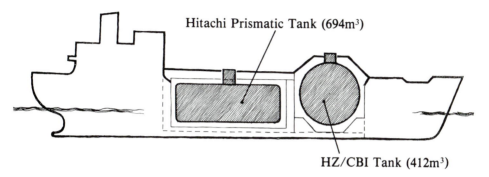

Fig. 13i. Hitachi-Zosen Prototype vessel *Sankyo Ethylene Maru*–1,100 m³.

By way of conclusion, an illustration is given above of the extent of the analysis felt necessary for designing the hull structure of one of these ships— more specifically to define the input with which to analyse the spherical tanks and their supports. This particular 'gross model' idealization relates to the Hitachi sphere – but all spherical designs are subjected to the same type of analysis, though in some cases the sphere covers are also included since they contribute to the overall strength, and particularly torsional rigidity, of the hull structure.

Idealization of the critical equatorial ring area is carried out in the manner shown; stresses in this area may also be examined by photo-elastic techniques.

The success of the spherical design insofar as ship orders are concerned has been impressive – although it has to be noted that such success has, to date, been confined to the Kvaerner-Moss design; no other spherical LNG ship has been built with the exception of the Technigaz, Hitachi and Sener prototypes – all of which were 'in-house' orders.

When presenting their design in 1970, Moss were able to quote orders/ options for four ships totalling nearly 240,000 m³; in 1973[5] there were 18 ships totalling 1,875,000 m³ built or on order, and the total in 1978 was 23 ships totalling 2,525,000 m³, of which 50 per cent is accounted for by the General Dynamics [7/8] shipyard – success indeed (see Chapter 16 for update).

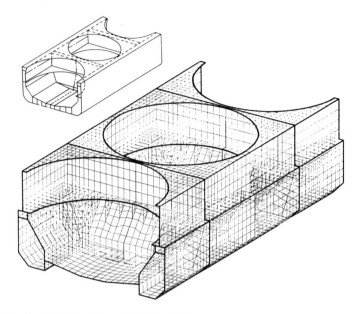

Fig. 13j.  Typical idealization of ships' hull.

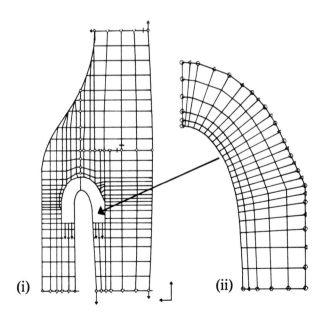

(i)                    (ii)

Fig. 13k.  Typical idealization of equatorial ring.

[1] 'The design of an 88,000 m³ LNG carrier with spherical cargo tanks and no secondary barrier' by R. Kvamsdal, H. Ramstad, R. Bognaes, H. J. Frank. LNG 2, Paris, 1970.

[2] 'Methane carriers with spherical tanks. The Technigaz Technique SSA – design and experience' by G. Bonnafus and F. Shaw. LNG 3, Washington, 1972.

[3] 'Pittsburgh-des-Moines/Gaz Transport's self-supporting system for series production' by S. Whitehead. LNG 73, London, 1973.

[4] 'Sener's LNG containment system – basic principles and design processes' by Dr J. Torroja Menendez. LNG 73, London, 1973.

[5] 'Experience from design, construction and initial commissioning of the first Moss Rosenberg LNG carrier – new developments' by R. Kvamsdal. LNG 73, London, 1973.

[6] 'Design of spherical shipborne LNG cargo tanks' by R. D. Glasfeld. SNAME, San Diego Section, 1975.

[7] 'Series Production of liquefied gas carriers' by P. Takis Veliotis. Gastech 78, Monte Carlo, 1978.

[8] 'A solution to the series production of aluminium LNG spheres' by P. Takis Veliotis. SNAME, Vol 85, 1977.

[9] 'The Hitachi Zosen/CBI spherical LNG tank ship system' by S. Murata et al and P.R. Johnson. Gastech 82, Paris.

# 14
# Market Prospects Generate Many New Designs

The success of the Algeria-UK project, coupled with high expectations of a rapid expansion of the market, encouraged many design groups to investigate new and improved (?) techniques for the shipment of LNG. Classification Societies were inundated with proposals for evaluation and requests for approval in principle during the late sixties and early seventies. Some of these proposals have been briefly mentioned in earlier pages, and many have been developed to an advanced stage, including the construction of prototype installations and seagoing tests; not all reached that stage, however, foundering for one reason or another on the drawing board or in the laboratory.

It is difficult to place these designs in strict chronological order because there was considerable overlap in their gestation and much of the work was carried out under conditions of close commercial security, only becoming public knowledge when they were formally presented at one of the annual gas or shipping conferences; in fact, it was only by attending such conferences that real progress of the technology could be followed and assessed – and opinions compared, either privately or in public.

The most significant of the many designs developed since the mid-sixties are therefore presented and discussed below, in the approximate order in which they were formally presented to the world at large.

## (1) Ocean Transport Pressure System (1963–8)[1]

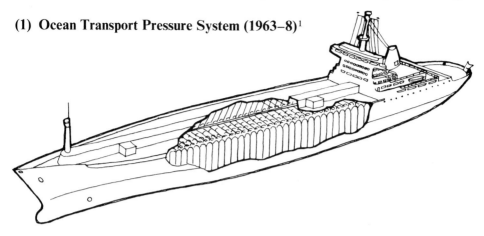

Fig. 14a. Ocean Transport ship design.

This group developed two LNG transportation concepts, both of which involved carrying the cargo at quite high pressures. The first was called the CNG (compressed natural gas) process, at 1,150 psig and $-75°$F; the second was the MLG (medium condition liquefied gas) process at 200 psig and $-175°$F.

This new approach was aimed at reducing the high cost of liquefaction which, although offset to some extent by a higher ship containment cost, would nevertheless result in a net overall saving in the delivered cost of gas.

Development work on both sides proceeded

> 'through the laboratory stage on to a pilot plant, subsequently through full-scale testing of equipment, systems and safety features and finally to a prototype programme involving a ship-to-shore facility combination.'

One particularly interesting aspect of the Ocean Transport Group's work was their approach to the bottle design which involved the adoption of

> 'a design philosophy that a bottle would leak before failure [and the use of] a new alloy, weldable, quenched and tempered steel with good strength – 11,000 psi – and good fracture toughness at operating conditions. Several manufacturers developed entries which were submitted to rigorous testing.'

The methods developed for testing the steels were later extended to aluminium.

The Group's final stage in their programme was to purchase and convert a Liberty ship, renamed *Sigalpha*, to carry either 820 MCF of MLG or 1,300 MCF of CNG. Several twenty-four hour demonstration voyages were made, with both cargoes, culminating in the receipt of ABS approval and USCG certificate of compliance for both systems.

Although technically proven, neither of the two concepts found 'buyers'— largely due to the high cost of the storage; attempts were made to develop a less expensive prestressed concrete shore storage unit but these clearly did not succeed in tipping the scales in favour of either system.

Despite this early set-back, the pressure concept – now further reduced to 60 psi – reappeared much later in the form of the Ocean Phoenix Transport design which is discussed later in the chapter (no. 9).

### (2) The McMullen System (1965–71)[2]

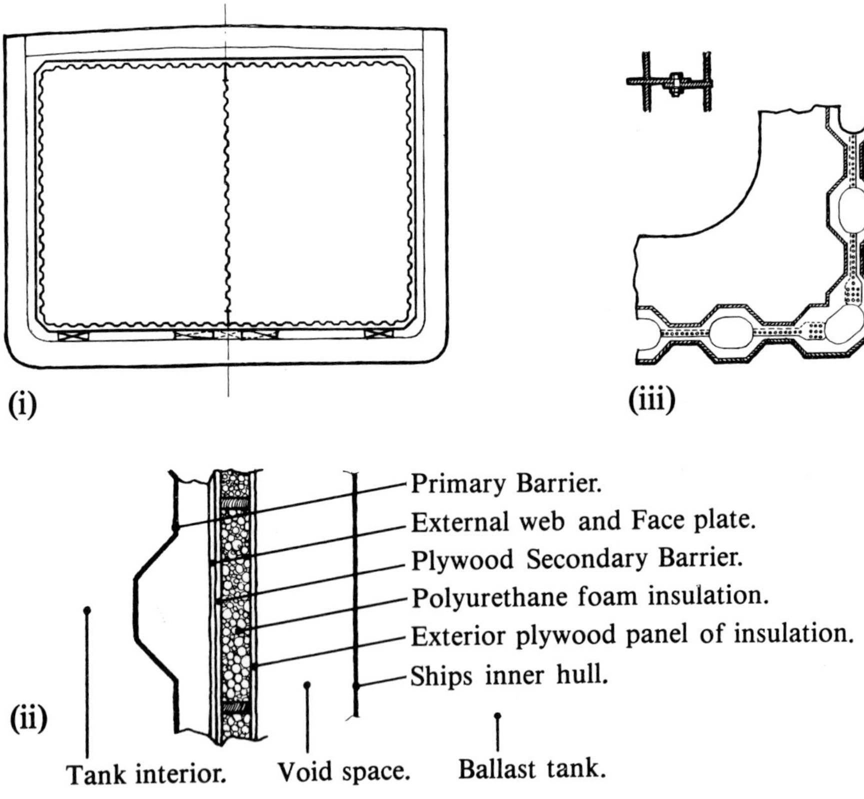

(i)                                                                (iii)

Primary Barrier.
External web and Face plate.
Plywood Secondary Barrier.
Polyurethane foam insulation.
Exterior plywood panel of insulation.
Ships inner hull.

(ii)

Tank interior.      Void space.      Ballast tank.

Fig. 14b. The McMullen System.

This was originally a double wall independent tank design, of either 9 per cent nickel or aluminium alloy, based on the corrugated bulkhead concept, familiar to tanker owners of the post-war period (fig. 14b[i]).

The main design problem associated with this concept was to devise a flexible connection between the walls to provide for differential contraction

164

and ship-induced relative movement which would not only function satisfactorily but could be properly assembled and inspected in service.

The design was subsequently modified around 1970 (fig. 14b[ii]) to one in which

> 'The web, or diaphragm, joining the primary and secondary barriers . . . has been converted into an external web and face plate and the outer corrugated skin deleted. A modular insulation system consisting of plywood-faced wooden boxes filled with polyurethane is mechanically attached to the tank external face plate structure. The interior plywood face of the insulation panels acts as a secondary barrier. The design of the primary and secondary barriers conform to the US Coast Guard requirements for an "independent, structurally semi-determinate, self supporting gravity tank". The upper portions of the secondary barrier serve to deflect any liquid leakage to the bottom while maintaining the space gastight. The lower portion of the secondary barrier is liquid-tight. Pumps are provided in the interbarrier space for removing leaked liquid to the primary tank.'

However, the design did not find favour. Possibly the early intertank leakage problems in conventional tankers with corrugated bulkheads was partly responsible, but at any rate it is doubtful if the design could have met the more stringent regulatory requirements which developed a few years later.

### (3) Semi-Membrane Designs (1961–72)
Although it may be recalled that the Chantiers de l'Atlantique shipyard developed and patented a semi-membrane in 1963, most of the real pioneering work on these designs was carried out in Japan.

The basic principle is simple, namely to construct the primary barrier of flat plating which may vary from 6 to 20 mm in thickness depending upon the

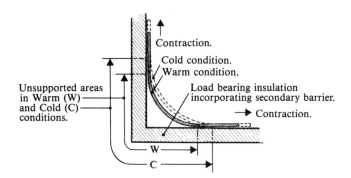

Fig. 14c. Principle of semi-membrane.

material and method of construction used; arrange for the flat tank walls to be uniformly supported by the insulation fitted on the inner hull/hold surfaces; also arrange for the corners to have a large radius and sufficient strength to be effectively self-supporting against the cargo load – being of cylindrical form at the edges and cylindrical at the corners; this latter is not too difficult. When the inevitable tank shrinkage occurs the unsupported area increases somewhat in extent, and changes in shape, but is still able to support the cargo loads.

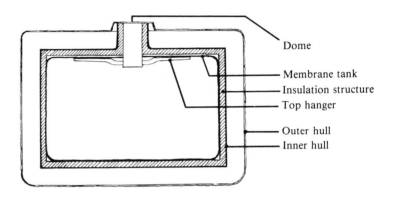

Fig. 14d. Bridgestone's membrane tank.

**The Bridgestone System (1972)**[3][4][5] (fig. 14d) has a relatively thin primary barrier of about 4 to 6 mm 9 per cent nickel steel, so that the top of the tank requires extra support hangers to prevent sagging at times when there is not positive pressure in the tank (during construction and at annual surveys). The original secondary barrier of plywood was later replaced by a 3 mm layer of stainless steel. Merits of the system claimed by the designers are:

> 'The method of manufacturing the integrated cargo tank outside of the ship permits easier fabrication of the oversized tank because appropriate jigs can be used, and welding and inspection can be done from both sides of the membrane. [Also] . . . The most outstanding advantage of this system is that the stress level of the entire primary container is known, being tensile stress only, and a design stress far lower than the allowable limit of the metal can be chosen.'

Two small ethylene tankers *Ethylene Dayspring* (1,100 m³) and *Ethylene Daystar* (890 m³) were built to the design in 1968 and appear to have operated satisfactorily in ethylene service since then – but no LNG ships have incorporated it as yet.

**The IHI System (1973)**[6] (fig. 14e), developed by the Ishikawajima-Harima

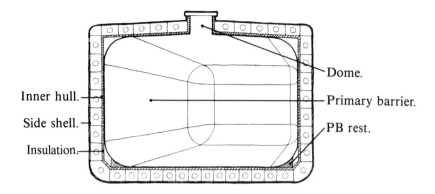

Fig. 14e. Cutaway of typical LNG tank.

Shipyard, differs from the Bridgestone system principally in its use of aluminium alloy (5083–0) rather than 9 per cent nickel steel; this results in substantially heavier scantlings (15–40 mm, depending on tank size) and renders each
tank completely self-supporting in the empty unpressurized condition. A
prototype tank of 6 m × 7 m × 8 m, about quarter full-scale, was constructed
in the shipyard in 1972 to evaluate fabrication procedures and behaviour in
the cold condition. In addition a 2 m × 2 m × 2 m model was subjected to
fatigue tests of $2 \times 10^6$ cycles at a pressure corresponding to the maximum
pressure expected once in twenty years.

The secondary barrier was plywood on a timber frame work which was
filled with foamed-in-situ polyurethane.

During the discussion at the 1973 Conference the Bridgestone representative commented:

> '. . . I am much astonished at the similarity of this system with our . . . system'

and another discusser:

> '. . . I am just wondering that if it had been considered in 9 per cent Ni steel and
> it had not been an inch thick, but something like 8 or 10 mm, then the only
> apparent difference between this design and the one we heard about [Bridge-
> stone] would have disappeared.'

## (4) The LGA-Zellentank System (1972)[7] [8]

This is an assembly of many interconnected tubular vessels arranged horizontally across the ship. The cylinders are standard machine made elements
which can be prefabricated in diameters up to 2.5 m and lengths up to 30 m or

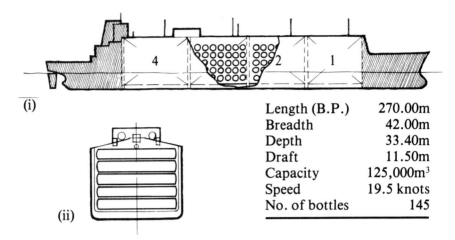

| | |
|---|---|
| Length (B.P.) | 270.00m |
| Breadth | 42.00m |
| Depth | 33.40m |
| Draft | 11.50m |
| Capacity | 125,000m$^3$ |
| Speed | 19.5 knots |
| No. of bottles | 145 |

Fig. 14f. LGA Zellentank design for 125,000 m$^3$ LNG ship.

more; they are spirally welded and supported at two points only, thus eliminating ship structure deflections; the design is based on pressure vessel rules, therefore no secondary barrier is required. Insulation is fitted to the hold inside surfaces. Due to low labour content and efficient use of material and ship volume, a 40 per cent saving in total containment cost was claimed by the designers in addition to such advantages as high safety factor and adaptability to hull shape.

Nevertheless the apparent complexity of the design and apprehension as to the tightness of the many interconnections between the cylinders has caused shipowners to shy away from the design.

### (5) The Linde Curved Wall Tank System (1968)[9]

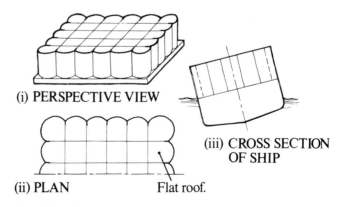

(i) PERSPECTIVE VIEW

(ii) PLAN     Flat roof.

(iii) CROSS SECTION OF SHIP

Fig. 14g. Linde-curved wall tank.

This design was proposed in 1968 as a marine application of a shore tank design; as such, considerable attention had been given to asymmetric loading due to roll and pitch – but little to the influence of hull structural deflections.

A flat roof was proposed with keys at top and bottom preferred as a means of location. Little has been heard of this proposal.

## (6) The Linde Multi-Vessels System (1974)[10/11]

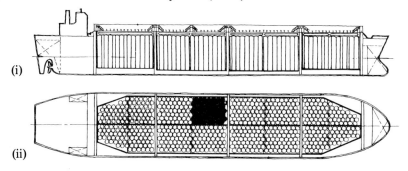

(i)

(ii)

No. of Holds  4.   No. of Tank Units  16.   No. of 'vessels'  694.

Fig. 14h.  Linde Multi-vessel-tank design for 125,000 m³ LNG ship.

This tends to suffer from the same handicaps as its predecessor in concept, the Ocean Transport Group system, and as such has not found favour. Considerable work was carried out on this design by its originators but once again shipowners tend to favour simple solutions if they can get them at the same price. However, it is interesting to compare Linde's thinking with that of Ocean Transport/Ocean Phoenix whose progression of thought has been diametrically opposite in direction but, it would appear, to better avail.

## (7) The Contranstor Proposal (1974)[12]

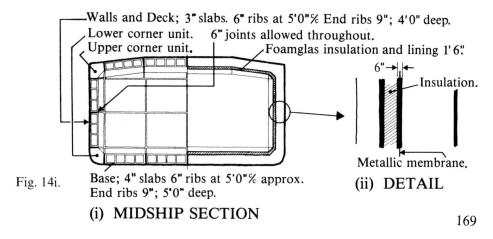

Walls and Deck; 3″ slabs. 6″ ribs at 5'0"% End ribs 9"; 4'0" deep.
Lower corner unit.    6″ joints allowed throughout.
Upper corner unit.        Foamglas insulation and lining 1'6".

6"→| |←

Insulation.

Metallic membrane.

Fig. 14i.    Base; 4" slabs 6" ribs at 5'0"% approx.        (ii) DETAIL
End ribs 9"; 5'0" deep.

(i) MIDSHIP SECTION

This disappeared from the scene almost as quietly and quickly as it appeared – in 1974; based on a concrete reinforced hull structure, it was more an attempt to provide a suitable 'vehicle' within which to fit any of the existing containment systems than to provide a new type of LNG ship.

It is recorded in this book only because its promoters were sufficiently serious to seek – and obtain – Lloyd's approval in principle for the design, but for one reason or another did not follow the project through.

### (8) 'Internal Insulation' Systems

Variously known as internal insulation, 'Wet Wall', Integrated or Integral, this genre of LNG containment began to appear in a definable form in the early 1970s. It had been well known that several of the major US aerospace groups, suffering from a severe cutback in the space programme, were diverting considerable manpower and funds into LNG containment, with a view to commercial exploitation of their recently acquired and highly sophisticated cryogenic technology and test facilities.

Their common objective was simple: to develop an insulation material, or system, which, being applied to the inside of the holds, could safely and reliably contain the LNG cargo without the need for metallic linings, aluminium tanks or other such costly items.

The idea was totally logical – it was, indeed, the 'ultimate containment system' – but the execution presented a few problems as Morrison had found twenty years earlier with his balsa system, which was after all precisely the same concept. Would the advanced technology which reached the moon, find a commercial solution to the transportation of natural gas by sea? Let us see.

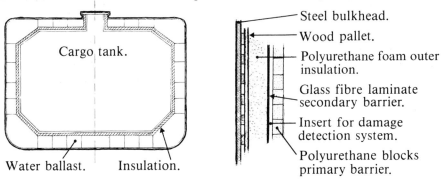

Fig. 14j. Midship section and insulation detail for Rockwell International Wet Wall Integrated Containment system.

**The Rockwell International 'Wet Wall' Design**[13] was a system based on polyurethane foam and glass fibre laminates. As can be seen in fig. 14j the system consists of two liquid-tight barriers and a layer of polyurethane foam insulation, mounted on a backing pallet of plywood. The inner, or primary, barrier is polyurethane foam blocks and the outer, or secondary, barrier is a fibrous glass reinforced plastic laminate membrane.

Factory made 'modules', constructed in this manner, are attached to the ship structure by studs, and joints between modules made tight to complete the integrity of the primary and secondary barriers. A detection system to monitor for leakage through the primary barrier is provided.

A prototype tank was built and tested and, because the system was never seriously marketed, one must assume that the Rockwell attempt failed to meet the required standard.

### The Owens-Corning (Perm-Bar II) Design:[14] this system

> 'consists of pre-engineered and factory-fabricated panels and splines [joining sections] [as shown in figs. 14k]. 'The panels are mechanically fastened to the ship's hull; both panels and splines are manufactured complete with all the barriers and insulating cores, allowing the total containment to be applied in one step.
>
> 'The panels (typically 6 ft × 9 ft) are of sandwich construction. The primary and secondary barriers are made from a specially developed glass-reinforced plastic laminate [FRP] which more than meets the stringent requirements of a cryogenic marine barrier.
>
> 'The core material is . . . polyurethane foam with appropriate honeycomb reinforcement.'

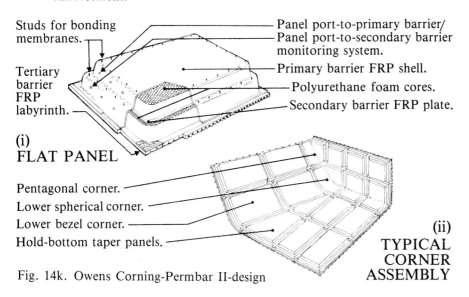

Studs for bonding membranes.

Tertiary barrier FRP labyrinth.

Panel port-to-primary barrier/
Panel port-to-secondary barrier monitoring system.
Primary barrier FRP shell.
Polyurethane foam cores.
Secondary barrier FRP plate.

(i)
FLAT PANEL

Pentagonal corner.
Lower spherical corner.
Lower bezel corner.
Hold-bottom taper panels.

(ii)
TYPICAL
CORNER
ASSEMBLY

Fig. 14k. Owens Corning-Permbar II-design

171

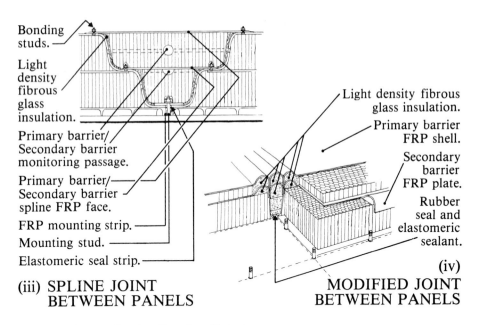

Bonding studs.

Light density fibrous glass insulation.

Primary barrier/ Secondary barrier monitoring passage.

Primary barrier/ Secondary barrier spline FRP face.

FRP mounting strip.

Mounting stud.

Elastomeric seal strip.

Light density fibrous glass insulation.

Primary barrier FRP shell.

Secondary barrier FRP plate.

Rubber seal and elastomeric sealant.

(iv)

### (iii) SPLINE JOINT BETWEEN PANELS

### MODIFIED JOINT BETWEEN PANELS

Fig. 14k. Owens Corning-Permbar II-design.

Some observers felt that there might be difficulty in achieving a satisfactory bond at the splines unless extreme care were taken in lining up the relatively inflexible panels, and so it appears to have proved, since a modification was subsequently made to this area (fig. 14k[iv]) which provides more scope for misalignment and overall assembly tolerances. By 1978 Owens-Corning had arrived at a satisfactory and economical geometry for the flexible joints – most importantly the interception cross – completed the structural and thermo analysis of the system and secured the US Coast Guard's approvals. However, by this time the LNG ship market potential did not justify the erection of a manufacturing facility so their marketing efforts were abruptly discontinued and the technology put into cold storage.

**The McDonnell Douglas Design**[15/16/17] is a

'bonded, three dimensionally (3-D) reinforced polyurethane insulation, originally developed for the interior of the upper stage liquid Hydrogen tank on the Saturn/Apollo moon rocket.'

Although originally designed and tested to provide a total containment system in itself, in its first practical application it was planned to be used in conjunction with the Gaz Transport membrane which would act as a primary barrier.

Fig. 14 l illustrates the principle adopted in the so-called GT/MDC design. The original proposal (i) was modified by the Sun Shipbuilding Company,

172

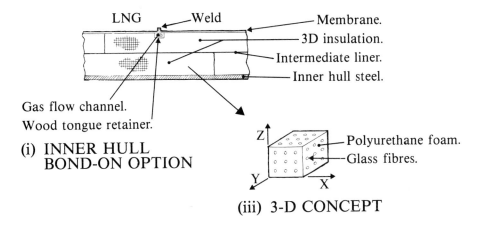

(i) INNER HULL
     BOND-ON OPTION

(iii) 3-D CONCEPT

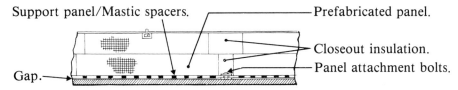

(ii) PREFABRICATED BOLT-ON PANEL OPTION

Fig. 14 1. McDonnell-Douglas 3D-foam system with GT membrane.

who were contracted to build two 125,000m³ ships to this design, in order to incorporate a 'stand off' capability (ii); this means supporting the system at a small distance from the inner hull by means of wood or hard setting mastic spacers to enable water leakage from the adjacent ballast spaces to be easily drained away and disposed of. A sensible precaution.

This short description hardly does justice to the fact that the McDonnell Douglas Company, who decided to enter the LNG arena as late as 1973, allocated a $20 \times 10^6$ budget and 100 men to the project at that time.

In the event these two ships were never built due to the downturn in the market.

**The Dytam Concrete LNG Design**[18] is, or was, because the promoters ceased marketing it in 1978, a radical departure in naval architecture in addition to being a new LNG ship design. The LNG containment was based

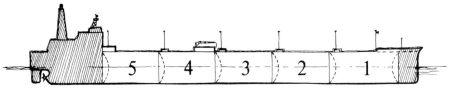

(i) PROFILE

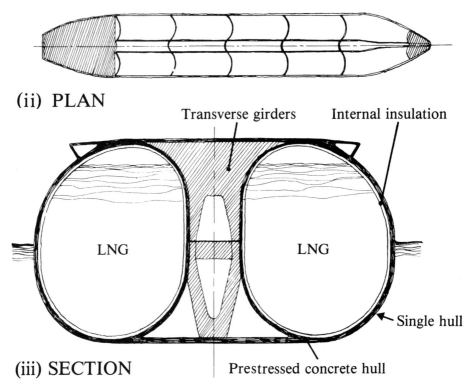

## (ii) PLAN

Transverse girders        Internal insulation

LNG        LNG

Single hull

## (iii) SECTION        Prestressed concrete hull

Fig. 14m. Concrete LNG carrier.

on the as yet untried Perm Bar II system–difficult, it could be imagined, to apply satisfactorily to the multicurvature of the hold walls. Certainly too much for any poor shipowner to swallow in one gulp. To quote a reaction from the discussion period when it was introduced at the Gastech 76 Conference in New York:

> '. . . it is very difficult for me to understand how we can sell to the public *a concrete vessel*–and you must remember that they believe that concrete won't float!–how we are going to sell them that and how we are going to sell them a non-double hull vessel.'

Spoken with much feeling by a member of the Pacific Lighting Corporation of Los Angeles who had been battling to introduce LNG into his gas-hungry State for three years at that time – and continued for a further three before giving up the struggle.

Dytam spent several million dollars developing their design from first principles and marketing it with great enthusiasm and conviction before finally admitting defeat in 1978.

## (9) The Ocean Phoenix Pressure LNG System[19]

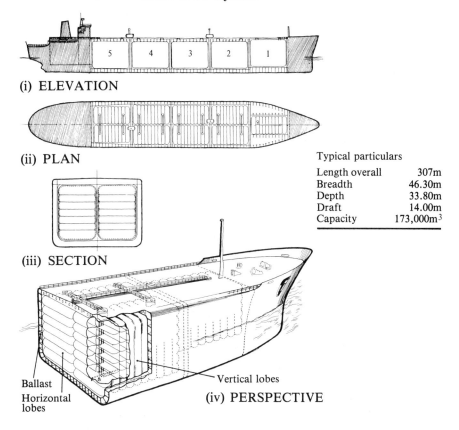

(i) ELEVATION

(ii) PLAN

(iii) SECTION

Typical particulars

| | |
|---|---|
| Length overall | 307m |
| Breadth | 46.30m |
| Depth | 33.80m |
| Draft | 14.00m |
| Capacity | 173,000m$^3$ |

Ballast

Horizontal lobes

Vertical lobes

(iv) PERSPECTIVE

Fig. 14n. Ocean Phoenix containment system.

This design was the reincarnation of the original Ocean Transport pressure concept[1] but now working at the much lower pressure of about 60psig and using cargo tanks of a much more economical design, namely a multi-lobe, trapezoidal type. Despite the apparent complexity, the tank was designed to meet IMO's Class 'C' (pressure vessel) category and thus required no secondary barrier, and largely because of its modular form it was claimed to be competitive in cost to more conventional designs.

The tanks could be constructed of 9 per cent nickel steel or aluminium alloy (5083-0); the insulation expanded mineral powder, filling the space around the tanks, or polyurethane foam blocks attached to the tanks, or inner hull. Locating keys are fitted at the top and bottom of each tank. The designers claimed that, by eliminating the final stage in the liquefaction process, and carrying both LNG and LPG as a single rich gas cargo, the delivered cost per BTU is significantly reduced–up to 20 per cent, for certain fields. Cost savings may also be achieved by the fact that the normal 'boil off' from the cargo is fully contained, simply by allowing pressure to build up in the tanks. Additional safety was also claimed by virtue of the tank compartmentation.

Classification approvals were obtained for both LNG and LPG applications following the successful (and quite exciting!) burst test of a 1:3 scale model of 90m³ capacity at the Harland & Wolff shipyard, N. Ireland in mid 1982. However, despite the energetic marketing of a 24,000m³ LPG design, the concept did not appear to offer sufficient technical or financial advantages to tempt owners to 'try out' the new system – particularly at a time when the market for such ships was still at a very low ebb.

## (10) The Verolme Vertical Cylindrical Design[20]

Plate 1. Verolme Cryogenic containment system.

This was originally introduced as a design for an extremely large (330,000 m³) LNG tanker, adopting the philosophy that a ship twice as big as any existing design must transport gas much more cheaply.

A vessel of this size would contain some 90 individual 'tanks' in its five insulated holds (18 per hold). The arrangement provides facility for easy inspection of both sides of the tanks which are located by their bottoms only thus leaving the tops free to 'float'. Cargo pumps, of relatively small capacity, are fitted in each of the 90 tanks.

As with the Ocean Transport system the design was claimed as being extra safe due to the compartmentation of the cargo, but largely due to the multiplicity of tanks, pumps and pipe connections, the design found no favour in the market place.

### (11) The Metastano Internal Insulation System[21]

This design was developed by the ASTANO shipyard in Spain, being announced in November 1978, and thus the most recent internal insulation concept. Once again, it was based on space programme insulation technology.

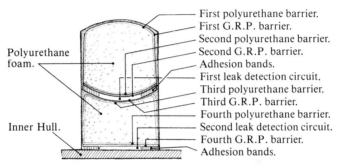

Polyurethane foam.

Inner Hull.

First polyurethane barrier.
First G.R.P. barrier.
Second polyurethane barrier.
Second G.R.P. barrier.
Adhesion bands.
First leak detection circuit.
Third polyurethane barrier.
Third G.R.P. barrier.
Fourth polyurethane barrier.
Second leak detection circuit.
Fourth G.R.P. barrier.
Adhesion bands.

(i) INTERNAL INSULATION SYSTEM
BARRIERS FOR L.N.G.

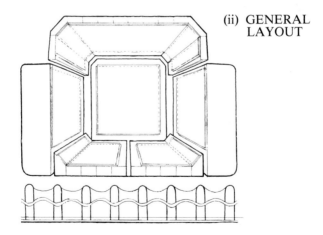

(ii) GENERAL
LAYOUT

Fig. 14o. Internal insulation system barriers for LNG.

As a design it is difficult to describe or portray in two dimensions; fig. 14o does not therefore add much clarification.

It is described by the designers as being

'. . . essentially composed of two [layers of] glass fibre reinforced plastic [GRP] boxes with covers either shaped or flat. Inside each box a basic cell is repeatedly reproduced in such a way that the cell's walls or partitions are tight and arranged as a chess board. The partitions are curved and the cell's surface is shaped or dome-like, alternatively changing its concavity.'

The illustration indicates that it is the yard's intention to fit most of the system to the prefabricated portions of the ship before final assembly on the berth, thus enabling most of this work to be carried out 'downhand' – likewise each 'panel' will be tested before final assembly.

## Summary

None of the designs described in this chapter have been adopted for a commercial LNG project. The nearest to success is, perhaps, the semi-membrane[3/6] to which design several fully refrigerated LPG ships have been built and are operating satisfactorily in that service.

In the first edition of this book (1979) the author suggested that although "some designs have died a 'natural death'...others are very much alive and only wait...for an upturn in the market, or a convinced shipbuilder or owner, or the coincidence of all three. Only time will tell of their success or failure."

In 1992 the market has finally 'upturned', but builders and owners remain unconvinced so, sadly, every one of these designs upon which so much dedicated effort and money were spent and midnight oil burned, seem destined to find only an honourable place in the history books.

[1] 'A new process for the transportation of natural gas' by R. J. Broeker. International LNG Conference, Chicago, 1968.

[2] 'LNG carriers: the current state of the art' by W. D. Thomas and A. H. Schwendtner (discussion). SNAME, Vol 79, 1971.

[3] 'Bridgestone's membrane tank' by K. Yamamoto. LNG Conference, Watergate Hotel, Washington, March 1972.

[4] 'Flat membrane type LNG carrier' by K. Yamamoto. LNG 3, Washington, 1972.

[5] 'Prototype tank test of semi-membrane LNG carrier' by E. Watanabe and K. Furuta. LNG 5, Dusseldorf, 1977.

[6] 'Development of the IHI flat tank containment system' by K. Kishimoto. LNG 73, London, 1973.

[7] 'The LGA-Zellentank system' by W. Kolb. LNG/LPG Conference, London, 1972.

[8] 'Further technical development of the LGA-Zellentank system and commercial considerations' by W. Wöber. LNG 73, London, 1973.

[9] 'Shell structure tanks for transport of LNG by ship' by E. Zellerer. International LNG Conference, Chicago, 1968.

[10] 'Calculation problems with multi-vessel systems' by E. Zellerer. Gastech 74, Amsterdam, 1974.

[11] 'Sea-transport of LNG in bottle-type vessels' by W. Foerg and R. Becker. LNG 3, Washington, 1972.

[12] 'Contranstor–an integrated system' by F. H. Turner. Gastech 74, Amsterdam, 1974.

[13] 'Whither the LNG ship?' by W. D. Thomas. RINA, 1975.

[14] 'Perm-Bar II–a new containment system for LNG carriers' by W. A. Swaney. Gastech 76, New York, 1976.

[15] 'Development of the 3-D containment systems' by Dr J. L. Waisman. Gastech 76, New York, 1976.

[16] 'Adapting a new containment system to a shipbuilder's product' by F. P. Eisenbiegler and J. D. Mazzei. Gastech 76, New York, 1976.

[17] 'A new Invar membrane containment system for construction of LNG carriers' A. Gilles and Dr J. L. Waisman. LNG 5, Dusseldorf, 1977.

[18] 'For LNG–a concrete answer' by A. E. Stanford, Dr K. Finsterwalder and W. C. Carvill. Gastech 76, New York, 1976.

[19] 'The Ocean Phoenix pressure LNG system' by E. K. Faridany, H. C. Secord, J. V. O'Brien and M. Banister. Gastech 76, New York, 1976.

[20] '330,000 m$^3$ Verolme LNG carrier' by C. Verolme and Dr A. K. Winkler. LNG 5, Dusseldorf, 1977.

[21] 'International insulation system "Mestano-20" for the storage and transportation of liquefied gases' by Dr E. Dominguez. Gastech 78, Monte Carlo, 1978.

# 15
# The IMO Code

The Intergovernmental Maritime Consultative Organization (IMCO) was officially established in 1958 as a specialized agency of the United Nations, solely concerned with maritime affairs. It became The International Maritime Organization (IMO) in May 1982. For convenience IMO will be used throughout this chapter, most of which in fact predates 1982.

## THE STAGE IS SET

By 1970 the El Paso Natural Gas Company had been able to state at the LNG 2 Conference in Paris:

'. . . the demand for gas in the United States shows no sign of abating.'

'. . . LNG technology is now proved. This is borne out by the fact that not only has Algerian LNG been delivered to the United Kingdom and France on a sustained basis since 1964, but other projects are now under way throughout the world. Alaskan gas is now flowing to Tokyo, Libyan gas will soon be moving to Italy and Spain [it started a few months later], and projects are now under construction to provide additional Algerian gas to France and to deliver gas from Brunei to Japan.'

'. . . LNG can be brought into the United States at acceptable prices . . .'

'Earlier this year, El Paso placed an order [in France] for the construction of two cryogenic tankers, which will be the largest of their type in the world, for transporting LNG from Algeria to the US. Each tanker will have a capacity of 120,000 m³. [This order was subsequently increased to three.] El Paso expects to place orders in the near future for additional tankers [six – which were placed in the USA for politico-financial reasons] providing for a fleet capable of carrying the one billion [10⁹] cubic feet of liquid gas per day . . .'

This was a period of intense activity – and fierce competition – amongst the designers, promoters and licensee builders; it involved both the existing LNG

180

containment systems, and the new systems under active development, all anxious to participate in the many orders about to be placed.

Proposals for improvements and refinements to the Conch, Gaz Transport and Technigaz designs were now being negotiated with the Classification Societies and the USCG who were already 'submerged' in proposals for new systems; the new spherical designs were in their final stages of development and approval–and rumblings were being heard about an impending breakthrough in the form of the internal insulation systems.

## THE CODE IS INITIATED[1]

It was in this atmosphere that representations were made to IMO, now in the final stages of their development of a Code for the construction and equipment of ships carrying chemicals in bulk, that the preparation of a similar Code for Gas Carriers was urgently needed. Quite apart from its general desirability it would provide considerable relief to the US Coast

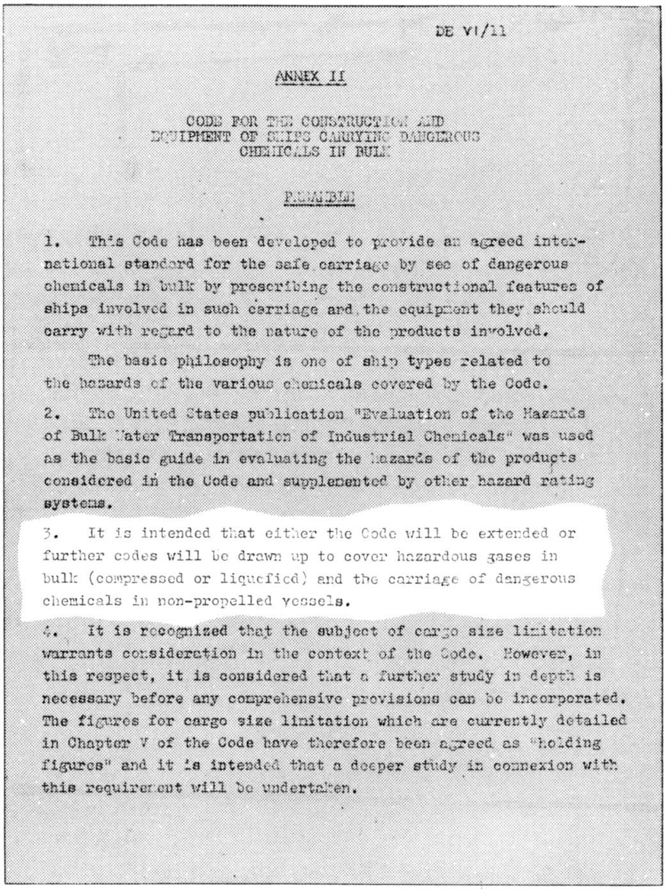

DE VI/11

ANNEX II

CODE FOR THE CONSTRUCTION AND
EQUIPMENT OF SHIPS CARRYING DANGEROUS
CHEMICALS IN BULK

PREAMBLE

1. This Code has been developed to provide an agreed international standard for the safe carriage by sea of dangerous chemicals in bulk by prescribing the constructional features of ships involved in such carriage and the equipment they should carry with regard to the nature of the products involved.

The basic philosophy is one of ship types related to the hazards of the various chemicals covered by the Code.

2. The United States publication "Evaluation of the Hazards of Bulk Water Transportation of Industrial Chemicals" was used as the basic guide in evaluating the hazards of the products considered in the Code and supplemented by other hazard rating systems.

3. It is intended that either the Code will be extended or further codes will be drawn up to cover hazardous gases in bulk (compressed or liquefied) and the carriage of dangerous chemicals in non-propelled vessels.

4. It is recognized that the subject of cargo size limitation warrants consideration in the context of the Code. However, in this respect, it is considered that a further study in depth is necessary before any comprehensive provisions can be incorporated. The figures for cargo size limitation which are currently detailed in Chapter V of the Code have therefore been agreed as "holding figures" and it is intended that a deeper study in connexion with this requirement will be undertaken.

Fig. 15a.

Guard in their now over-burdening 'Letter of Compliance' commitments under which they were required, by law, to inspect every gas carrier (LNG or LPG) desirous of entering a US port and to provide a 'letter of compliance' to that ship, so satisfying the authorities as to its standard of design and construction and thereby ensuring the safety of the port.

In September 1971, IMO held the first meeting of the Ad Hoc Working Group of the Subcommittee on Ship Design and Equipment at which fifteen countries were represented, together with 'observers' from four related associations. Their initial task was to define the general format and scope of the Code, and to allocate responsibilities for formulating a first preliminary outline on which subsequent discussions could be based; and also, most important, to define when, and to what ships, it should apply. It would be designed to cover both LNG and LPG cargoes, fully refrigerated, part pressure/part refrigerated and fully pressurized. A daunting task.

## OBJECTIVES OF THE CODE

In their paper to the Society of Naval Architects and Marine Engineers of New York in 1977,[5] three members of the US delegation (including the Chairman) summarized the basic philosophy behind the Code as follows:

'In developing the Code, the major concern was the possibility of a cargo release posing a hazard to a wide area. While most liquefied gases do not pose a water pollution threat, other potential hazards such as flammability, toxicity and the extreme low temperature of carriage call for special attention to cargo containment under both normal and emergency conditions. The central theme or philosophy of the Code is to provide maximum attention to cargo containment and to minimize the release of cargo in the event of a casualty. The code has, therefore, been based upon sound naval architecture and engineering principles and the best understanding of gas technology today. The Code emphasizes ship design and equipment; however, it also recognizes that in order to ensure the safe transport of liquefied gases, the total system must be appraised. Other equally important facets of the system such as operations, traffic control and handling in port remain primarily the responsibility of the individual governments where the vessels trade.'

There were a number of issues which required early resolution before any progress could be made.

    (i)   *Definitions*: it quickly

'became obvious that commonly used terms in the marine [LNG] industry had very different meanings in various parts of the world. Therefore special efforts were made . . . to develop agreed definitions. That users of the Code thoroughly understand the definition section cannot be overemphasized, as in many instances the definitions of terms set the limits of requirements.'

182

(ii) *Application*: would the Code only apply to new ships, or existing ships, or both? At what point is a ship new or existing? This was resolved by defining the date of building contract, 'keel laying' and delivery dates; with a similar definition for major conversions.

(iii) *Types of containment*: these were broken down into four major categories (a) Integral, (b) Membrane, (c) Semi-membrane and (d) Independent, Types A, B & C. This took into account all 'existing' designs and seemed to allow for all foreseeable developments.

(iv) *Code review*: in the light of the fact that the technology was still developing quite rapidly, and not wishing to place any restriction on such development, it was agreed that machinery for a regular review of the Code would be established.

Although not part of the terms of reference of the working group, IMO encouraged regular progress reports back to industry via, for example, ICS representation or by formal papers to International Conferences, thus enabling owners, builders and designers to anticipate the final Code requirements as far as possible. This policy was an invaluable aid to all those involved in the industry in enabling them to absorb such changes as were to become mandatory with as little additional extra cost as possible – an important factor on what was then a $120,000,000 investment.

The work in Committee could not have been achieved without considerable back-up support for the delegates themselves; and establishing this essential infrastructure was a major organizational task in itself. For example in the USA:

'In order to more effectively participate in the development of the Code, the Coast Guard formed a special task group under the Chemical Transportation Industry Advisory Committee [CTIAC]. Members of this task group were expert representatives of the American Gas Association, the Compressed Gas Association, the American Petroleum Institute, the American Institute of Merchant Shipping, the American Waterway Operators, the American Bureau of Shipping, the Shipbuilders Council of America, and the Society of Naval Architects and Marine Engineers. This task group met frequently with the Coast Guard to develop and review the emerging Code, and several of its members have participated along with Coast Guard personnel at IMO meetings. The project was approached by the task group by first developing drafts of revised US regulations for gas ships, then comparing those drafts with proposals from other countries. The method proved to be very effective'.

In the UK, the Chamber of Shipping arranged regular meetings of the International Chamber of Shipping, and also meetings to brief the Department of Trade of the UK shipowners' position. Similar arrangements were made in all participating countries. Perhaps the US arrangements were the best organized–certainly their presence was felt the most by the Working Group and their points both forcefully and competently argued.

There is little doubt that by adopting these methods the experience of almost every expert in the LNG and LPG field, worldwide, was directly or indirectly fed into the Code and its progress was widely disseminated.

## THE MEETINGS

Meetings of the Ad Hoc working group were held at regular intervals between the first, at IMO Headquarters, in September 1971 and the last, also in London, in 1975—it thus took four years to complete. Throughout this period the group was guided by its Chairman, R. J. (Bob) Lakey, with a blend of Texan patience, good humour, tact, loss of patience combined with gentle, but at times not such gentle, pressure.

The numerous delegates and observers took up prepared positions, lost ground, gained ground, prepared meticulously, argued strongly. No doubt this is standard procedure in international conferences–but it was a privilege to participate.

The basic procedure was for groups to prepare drafts for certain chapters between meetings. This was the 'homework'; these would be discussed in full session and subsequently revised. Small working sub-groups would be 'despatched' to a quiet room, with instructions not to return to the main meeting until their redraft was complete.

**At the second meeting**, which the USCG hosted in Washington, May 1972, it quickly became apparent that the only way in which the USA and the USCG in particular, would be able to accept the final Code was if the Chapters on cargo containment (IV), cargo handling (V) and materials of construction (VI), were prepared and agreed by all the major ship classification societies, i.e. by IACS (the International Association of Classification Societies).

Having accepted the challenge, the Classification Societies, hitherto in disagreement on a multitude of items, initiated a crash programme within IACS and, indeed, produced an agreed position on this absolutely vital section of the Code. As Bob Lakey reported some three years later[2]

> 'Without the assistance of IACS, the IMO Working Group would have been unable to complete its task.'

**The meeting in Gdansk, Poland,** in May 1974 was memorable, not so much for the hard work–this was now becoming normal–but for the wonderful

Plate 1. — Participants at the second meeting, in Washington.

*Standing (L to R).* 1. Dr A. Basso (Italy)  2. Mr G. Stubberud (Norway)  3. Mr M. Oshima (Japan)  4. Mr T. Mano (Japan)  5. Vice-Adm. W.F. Rea, USCG (USA) Host  6. M.J. Engerrand (IACS)  7. Mr Arne Bang (Norway)  8. Mr R.J.C. Dobson (UK)  9. Capt. (now Rear-Adm.) H.H. Bell, USCG (USA)  10. Mr R.K. Gregg (USA)  11. Mr T. Powers (USA).
*Seated (L to R).* 1. M. J-P. Christophe (France)  2. Mr K.G. Doehrn (Fed. Rep. of Germany)  3. Mr R. Graham (Canada)  4. Mr R.J. Lakey, USCG (Chairman)  5. Adm. H.C. Shepheard, USCG Retd. (ICS)  6. Mr R.C. Ffooks (ICS)  7. Mr Bakke (Norway)  8. Mr R.K. Roberts (UK)

hospitality extended to the Group, not forgetting the compliment paid it by a request that senior members of their universities and industry might listen to the words of wisdom being exchanged across the table.

Evenings were spent visiting local architectural treasures, of which there were many, and attending a special Chopin recital. After these 'formalities', which were splendid in themselves, and the vodka, informal discussions in the hotel rooms were 'de rigueur'; these rooms, being less pretentious in their facilities than most western style conference hotels, lent extra atmosphere to the occasions and many 'barriers' were broken.

185

Plate 2. In session.

Plate 3. Attentive listeners

Plate 4. Chopin recital.

Plates 2–4. Gdansk-Poland May 1974.

One such evening provoked an incident which became an 'in-joke' and can now be more widely shared. Having on two occasions drawn attention to the difficulties which might be experienced by shipowners to certain proposed items in the Code–to cries, from the assembled government representatives, of 'Oh-poor shipowners', the ICS delegate decided to pass round a 'hat' for contributions — duly given in the form of small coins of mixed denominations, bottle tops and buttons. On his return to the UK the same representative felt obliged to thank the Group formally through its Chairman for its generosity.

INTERNATIONAL CHAMBER OF SHIPPING

GRAMS : LOGBOARD, LONDON E-C 3
PHONE : 01 - 283 2922.
TELEX : 884008.

30-32 ST. MARY AXE,
LONDON, EC3A 8ET

24th June, 1974.

Robert Lakey, Esquire,
Chairman,
Ad Hoc Working Group - Gas Carriers,
IMCO Headquarters,
LONDON.

Dear Mr. Lakey,

    The International Chamber of Shipping wish to acknowledge with gratitude receipt of the donation* presented to their representative Mr. Roger Ffooks by your Working Group.

    In order to conform with the spirit in which this gift was volunteered, it has been unanimously decided to inaugurate a fund, to be known as the BALTYK TRIPLE-CUSHION FUND which, as its name implies, will be used for the relief of shipowners to whom the IMCO Gas Carrier Code causes undue suffering, particularly in regard to the physical discomfort, loss of sleep and general privation of their senior executives.

    You may be interested to know that at the first meeting of the B.T-C. Fund it was unanimously decided that as an expression of the sincerity of its purpose an immediate, though modest, "ex gratia" donation should be made to Conch Methane Services in recognition of the special aggravations, harassments and provocations to which this small but tenacious company has for so long been subjected. It is with deep satisfaction that we have learned that this grant has already produced a significant, albeit temporary, relief to the executives of this company.

    In conclusion, the Chairman and officers of the B.T-C. Fund request that you thank all those who contributed to this gift, especially the U.S. and Japanese delegations, and would like to express once again to you sir, their deep gratitude for this generous and spontaneous gesture.

Yours most sincerely,

R.C. FFOOKS
(I.C.S. Representative)

*(£ 9.00; $0.13; ¥ 1.00)

Fig. 15b.

187

The 'triple cushion' was inspired by the 'Baltyk' hotel beds–on which one lay on three separate 'cushions'.

All in all, considerable progress was made in Poland.

**The next meeting was Hamburg,** in October 1974; the Code was nearing completion and the final rearguard positions were being fought. The difficult question of filling limits in relation to venting (liquid or vapour) had been disposed of in Poland; now it was the Americans clinging to their requirement for special steel grades in the ship's deck and outer shell–Classification Society territory! And also stress factors for Class C tanks. In the end it was necessary to agree to differ on both these items.

The IACS group were working into each night, a second group was formalizing special requirements for Chapter XVII; despite the local attractions there was no spare time at all at this meeting. But in the final analysis the Chairman was able to announce, when introducing his 'progress report' at the Gastech 74 Conference[2].

> 'Last July [two months after Gdansk] when I was asked to give this paper, I accepted with some trepidation, because I fully realized that approximately three weeks before this conference the IMO Working Group would be meeting at Hamburg to complete the drafting of the IMO Code for Liquefied Gas Tankers. I asked myself the question, would I have good news, or bad? I think if you had polled the members of my Working Group, the large majority would have said I had bad news today. I expect the answer depends a bit on your point of view, but I feel I have good news. The Working Group virtually completed its drafting of the Code in Hamburg. The fact that it is virtually completed is a tribute to the members for their dedication and willingness to work five very long days in Hamburg.'

Not hitherto recorded is the fact that on the last day of the Hamburg session the 'hat' went round again for the Baltyk Triple Cushion Fund–this time for the relief of the exhausted IACS group!

**THE CODE ITSELF***

In November 1975 the Ninth Assembly of the Inter-Governmental Maritime Organization (IMO) adopted the 'Code for the Construction and Equipment of Ships Carrying Liquefied Gases in Bulk'; IMO Resolution A.328 (IX); and recommended that Governments incorporate its contents into national regulations as soon as possible.

* *Note:* Reference [5] is essential reading for the student of the Codes.

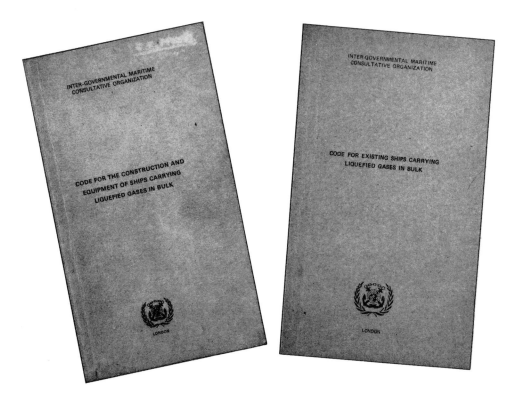

Fig. 15c.

Well before the Code for new ships had been completed, it became clear that a separate Code would be required for existing ships; technological advance had been so rapid that, although many ships in service – or soon to enter service, could be by no means deemed unsafe, at the same time they did not, and could not, meet the new Code in every respect. They were thus liable to be 'banned' from many ports at some future date – an unrealistic and untenable situation.

Within twelve months a Code for existing ships was produced following very closely the contents and format of the new ship Code. It took cognisance of the difficulties of bringing existing ships into line and considered each change on its merits, allowing varying periods of time to effect such changes as were deemed necessary. As in the case of the new ship Code, machinery was established for its review from time to time. It is worth noting that

> 'The US objective for the Code for Existing Ships was not completely attained and, therefore, the Code cannot be fully adopted. Nevertheless, it does provide a forward step towards maritime safety as it does provide port authorities a common basis upon which they can review vessels in the future.'[5]

As to Code review, in answer to a question at Gastech 75[3] on this point, Lakey replied

> 'Yes, machinery has been set up within IMO to do the task you just mentioned as well as several other jobs. Beginning in May [1976] there will be a new sub-committee on Chemicals. Its terms of reference and its work programme include the maintenance of the various Codes we have worked on in the past six or seven years. And the schedule is for that sub-committee to meet three times in every two year period. So that there will be ample and adequate opportunity to keep the Code under constant review as we all wanted in the beginning.

Regular reviews of the Code have been held since it was first published with very few significant modifications – which in itself bears testimony to the thoroughness of the original document.

One major amendment was the introduction, in 1979, of a fifth cargo containment tank type, namely the 'internal insulation tank'. This was introduced to accommodate Shell's sprayed-on polyurethane foam system, developed for LPG cargoes, but, one suspects, with an eye to LNG at some future date. The system consists of foam sprayed onto the inner hull (vide the Conch insulation cum secondary barrier system for LNG – which failed, p200) and provides both the primary cargo containment and insulation. The fact that the Shell system was not entirely successful in service and was subsequently 'withdrawn' does not preclude the possibility of future success; at any rate such a system is now provided for in the Code.

**To summarize:** when IMO announced, and rightly so, 'Let there be a Code', two were produced; and it is the opinion of at least one participant in their making that, while they may not be perfect, it is difficult to see how better, or more realistic, documents could have been written. That they were comprehensive and deeply thought out is indisputable; that they were universally popular is less certain, particularly in respect of their effect on smaller LPG ships, many of which became prematurely obsolescent as a result.

Many lasting friendships, but few enemies, were made by those who were directly and indirectly responsible for the production of the IMO Code. Much was learned and much was achieved, and that, at least, is a worthwhile contribution to the art – or science – of Natural Gas by Sea.

[1] Report to the Maritime Safety Committee by the Sub-Committee on ship design and equipment DE VI/II.

[2] 'The IMCO Code for liquefied gas tankers' by R. J. Lakey. Gastech 74, Amsterdam, 1974.

[3] 'The IMCO Code for gas tankers. A review of the finalized Code' by R. J. Lakey. Gastech 75, Paris, 1975.

[4] 'New regulations for liquefied gas carriers' by A. E. Henn, and T. R. Dickey. Gastech 75, Paris, 1975.

[5] 'A review of the IMCO Code for gas ships' by J. W. Kime, R. J. Lakey and T. R. Dickey. SNAME, San Francisco, 1977.

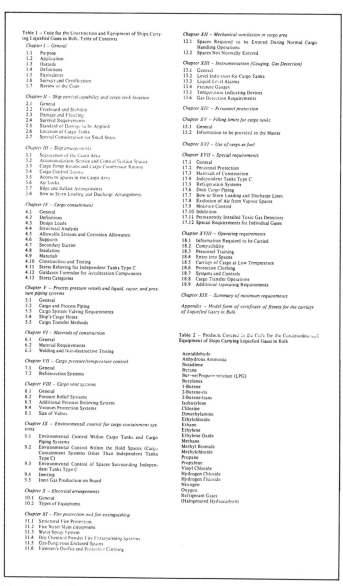

Fig. 15d. The original IMO Code: Contents.

# 16
# A Decade of Consolidation

The period between 1979 and 1990 may be considered one which proved without any doubt the reliability and safety of LNG transportation by sea. But it was not without its dramatic events and technological interest as this chapter will describe.

Japanese shipyards entered the field of LNG shipbuilding in their usual efficient manner and had built up a dominating position by the end of the decade. The great El Paso project of 9 x 125,000m³ ships, built to import gas to the U.S.A. from Algeria, collapsed in 1980 − no fault of El Paso; and perhaps the most tragic, and fascinating, of all was the virtual disintegration on trials in June 1979 of the insulation/secondary barrier system of the three Conch-designed ships, built by the Avondale Shipyard for this same project. This dramatic event, with all its repercussions, both financial and technical, was inevitably soon followed by the demise of the Conch company itself − but not, happily, by the disappearance of the trapezoidal tank concept which they had pioneered; the active development of its successor was already in the capable hands of the redoubtable Dr. Fujitani of the Ishikawajima-Harima Shipyard in Japan.

During this same period two new major projects were successfully mounted, but this chapter is concerned with recent developments in ship technology and its safety − always a matter of deep concern, and last but no means least the Conferences − which, by attending, the keen observer is able to keep in close touch with all aspects of LNG transportation and storage as they develop.

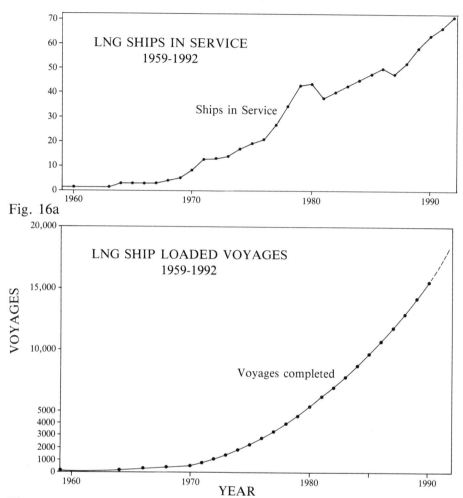

Fig. 16a

Fig 16b

## CONTAINMENT SYSTEMS UPDATE

By the end of 1991 each of the three main containment systems had clocked up well over 3,000 round trips[8] — the actual loaded voyages being:

| | |
|---|---|
| Technigaz Membrane .. .. .. .. .. .. | 3,570 |
| Gaz Transport Membrane.. .. .. .. .. | 5,183 |
| Moss Rosenberg Spherical.. .. .. .. .. | 4,358 |

Each have, as a result of their service experience, undergone relatively minor modifications a) to improve their reliability and potential life span, and b) to reduce their daily cargo boil off rate by improvements in their thermal efficiency.

Common to all systems has been the recent move to reduce capital and operating costs by reducing the number of cargo tanks from five to four in the standard 125,000m³ size vessel.

## THE TECHNIGAZ MEMBRANE

The original Technigaz membrane 'TGZ MkI' comprised a 1.2mm thick cross-corrugated 304-L grade stainless steel primary barrier (plate3, p.129) supported on a plywood faced balsa panel insulating/secondary barrier system as has been described in detail in Chapter 11. It has been installed in a total of 14 ships including five of the seven 75,000m³ ships employed on the Brunei-Japan run since 1971, and four 125,000m³ ships. Although three of the latter, having been ordered by El Paso, have seen little service, no in-service problems have been experienced with this system and no leakage of cargo through the primary barrier has ever been reported.

The only incident of note in a trouble-free service to date has been an overflow during loading which caused quite severe deck cracking – the repair of which is fully described in Ref[1]. During this repair the opportunity was taken to check the mechanical properties of the insulation/secondary barrier components after 16 years service. No deterioration was found – a comforting discovery!

The new Technigaz Mk III system (Fig. 16c), also briefly mentioned in Chapter 11, is being installed in an 18,800m³ vessel due to enter service in 1993.

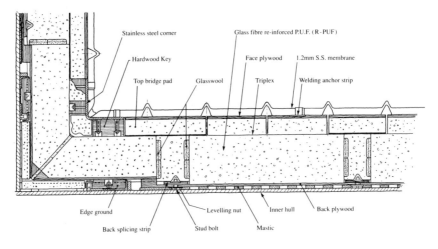

Fig. 16c. Technigaz Mk III. membrane — cross section at corner.

In common with other containment systems Technigaz have plans available for a four tank, 125,000m³ ship.

It should be noted that the membrane itself has remained unchanged since its introduction in 1971, although since that date significant strides have been made in automatic welding techniques which can now be achieved for the majority of the surface area.

## THE GAZ TRANSPORT MEMBRANE

The Gaz Transport membrane, also described in detail in Chapter 11, has more 'ship years' experience behind it than the Technigaz system but did develop problems as it moved up into the 125,000m³ size vessels, the rectifying of which required the affected ships to be taken out of service for repair; laid up vessels have been modified before reactivation.

There were two main problems; the first, already briefly referred to in Chapter 11, was the localised crushing of the insulation boxes in areas of high loading, due to cargo sloshing. This was easily dealt with by increasing the ply thickness and strength of internal dividers in the areas so affected. The second was leakage in the secondary barrier due to fatigue cracks developing at the points where the primary barrier corner angle tie rods passed through the secondary barrier. The 'flexible' connections at these locations began to fail after a few years service (again only in areas of high loading) and although various redesigns have been successfully adopted for existing ships, the problem has been eliminated in all new ships by replacing the tie rod (point load) arrangement by a continuous support system – nearly always preferable in ship construction. This new design is also claimed to have a lower heat leak at the corners.

Natural Gas by Sea

To alleviate sloshing loads in the larger tanks now being built, increased top chamfers are being reintroduced; compare the midship section of *Polar Alaska/Arctic Tokyo* (Fig. 11i) with (Fig. 14j) which latter closely resembles the section of the 1980s vintage designs. The large chamfer also eliminates the tank filling restrictions placed on membrane designs to control sloshing forces, Ref[4].

The Gaz Transport membrane has to date been installed in some 22 ships, 15 being in the 125/135,000m³ range; additionally 5 x 130,000m³, four tank, ships are on order – these will incorporate the large top chamfer and new box-type corner restraints.

Except as already mentioned, no LNG cargo leak has been reported in the GT primary barrier and the resilience of the system as a whole was demonstrated for all to see when *El Paso Paul Kaiser* grounded at speed off Gibraltar in 1979 (see p206 and Refs[4/5/7]). We do not need any more such examples to prove the point!

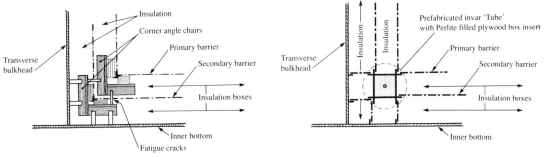

ORIGINAL CORNER DESIGN          NEW CORNER DESIGN

## THE MOSS SPHERES

We left the Moss spheres in Chapter 13 with 23 ships totalling 2,525,000m³ built or on order − this was 1979; the five 126,300m³ U.S. built ships having entered service during 1977-8 on the Indonesia-Japan run. The number at mid 1992 stood at 45 ships totalling 5,449,700m³ (33 in service, 12 on order).

Despite numerous uncomplimentary comments from the 'competition' about their outward appearance, and a flurry of rumours from time to time relating to the integrity of the bimetallic joints, all these ships have operated safely and reliably since they entered service.

The first four-tank 125,000m³ ships, built by Mitsubishi Heavy Industries, entered service on the Australian North-West Shelf to Japan project in 1989.

Apart from the increase in tank capacity from 25,000m³ to 31,875m³ moving from the five tank ship to four tanks, which has meant a 9% increase in the tank diameter, the only significant change from the earlier designs has been in the skirt construction.

The original design incorporated a single transition joint between the upper, aluminium, section and the lower carbon steel part. More recent designs, which include the four tank 125,000m³ ships, have introduced a stainless steel "thermal brake" between the aluminium and carbon steel elements of the skirt. The effectiveness of stainless steel as a thermal brake can be judged by comparing the thermal conductivities of the three materials involved. (Fig 16e).

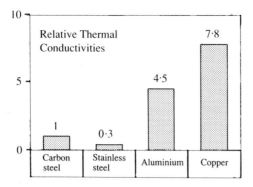

Fig. 16e. Comparative thermal conductivity.

Details of a typical present day arrangement are shown in Fig. 16f from which it will be seen that a neater design has been introduced for the bottom location of the central column through which pass cargo piping, instrumentation and personnel access to the tank.

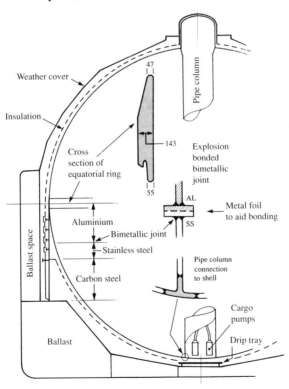

Fig. 16f. Cross section of Kvaerner-Moss spherical tank and support arrangement.

197

## TRAPEZOIDAL, SELF-SUPPORTING TANKS AND IHI-SPB

Following the successful entry into service of their first two ships in 1964, Conch began to look for a less expensive insulation/secondary barrier system than the balsa panels which, though effective, was very labour intensive and threatened to price their design out of the market. Two avenues were investigated: (1) a plywood/glass fibre splash barrier at the sides and a balsa tray at the bottom − this concept required that the integrity of the tanks could be proven and classification agreement obtained (not so easy in the 1960's) and (2) the substitution of balsa panels by a sprayed-on polyurethane foam (PUF) insulation system. After a number of preliminary 'screening' tests of both designs, the second was selected as the most cost effective and practical route to pursue. In fact, not all the balsa was eliminated; rows of balsa panels were retained on the bottom as tank supports and keyways; balsa was also retained in all corners and at the hold chamfers and top edges, forming a 'picture frame' of panels within which the PUF was restrained.

The main flat areas were intended to be automatically sprayed with twelve ½″ layers of foam, with a nylon netting between the two final layers to act as a crack arrester. On the surface of the foam was bonded 6″ of glass fibre, to ensure that the foam surface operated at a temperature well above − 165°C. This same combination of materials was installed between the bottom support panels. Fig. 16g shows a cross section of the overall insulation layout. Plate 1, shows the cargo tanks under construction.

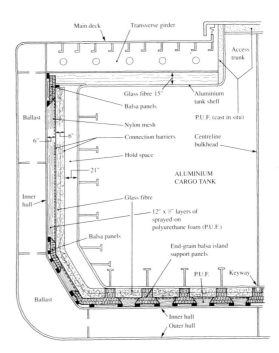

Fig. 16g.  Conch design for 125,000m³ ships built for El Paso.

198

Plate 1.   Conch tanks for 125,000m³ ships under construction at Mobile, Alabama — 1978.

199

The system was extensively 'prototype tested' by fatigue testing a number of 8' x 4' panels designed to simulate full scale installation, normal operating and emergency conditions. Each panel was tested to the equivalent number of cycles which would be experienced in four 20 year ship lives. All necessary approvals were obtained.

Orders for three 125,000m³ ships incorporating this latest Conch design were placed by El Paso with the Avondale Shipyard, U.S.A. in May 1973 for delivery in 1976/7. The yard chose to contract out the construction of both tanks and insulation which latter was, in any case, of a proprietary formulation.

It is stated above that it was *intended* that the foam should be applied by automatic equipment under carefully controlled atmospheric conditions, this latter being an essential requirement for its successful application and interlayer bonding; but as construction progressed the foam manufacturer and contractor were unable to develop satisfactory automatic spray equipment, and it eventually became necessary to apply the entire surface of each of the five holds of each of the three ships by hand − a formidable task.

The first of the three ships was completed and began cold trials in May 1979. After precooling the tanks down to cargo temperature, and holding this condition for several hours for temperatures to stabilize, a number of inner hull temperature gauge readings began to fall to abnormally low levels. Trials were immediately stopped. Upon return to the shipyard for examination the PUF system was found to be extensively cracked and delaminated; the crack arresting nylon mesh had not arrested cracks, the interlayer bonding had failed widely. It was, in effect, a total disintegration of the system.

What went wrong and why? Was the prototype fatigue testing inadequate? Was the quality control during installation inadequate? Was the formulation modified at some stage to accommodate large scale commercial application? Did unforeseen convection take place between the glass fibre and PUF? To date no satisfactory explanation has been put forward − but of one thing there is no doubt, the failure was total and the ships an effective write-off as LNG carriers − and Conch was wound up by its shareholders shortly thereafter. A sad end to the trapezoidal tank design − but was it?  Read on.

In 1965 Conch had sought tenders from Japanese shipyards for two 40,000m³ ships for the Nigeria-U.K. project then under study. The Ishikawajima Harima yard had sent a team of engineers to London to more fully understand the technology for which they were bidding. Subsequently they, and other Japanese yards, had built a number of fully refrigerated LPG ships and small ethylene tankers utilizing the trapezoidal tank design.

By 1982, confident in their full understanding of the technology and with experience of building over 20 LPG/ethylene ships with trapezoidal cargo tanks – all with proven satisfactory performance, and now having the benefit of vastly improved computerised finite element analysis techniques available – IHI formally entered the LNG field at the Gastech 82 conference in Paris with their own IHI-SPB design.

Though superficially very similar to the Conch design, it had two important differences:

(1) The cargo tanks were to be designed and built to meet IMO's 'leak before failure' criteria, i.e. they would be Independent tanks type B, which are "tanks designed using model tests, refined analytical tools and analysis methods to determine stress levels, fatigue life and crack propagation characteristics".

(2) The insulation would be attached to the tanks, *not* to the inner hull; and because the tanks had been upgraded from Type A to Type B a full secondary barrier was no longer required – just 'small leak protection'.

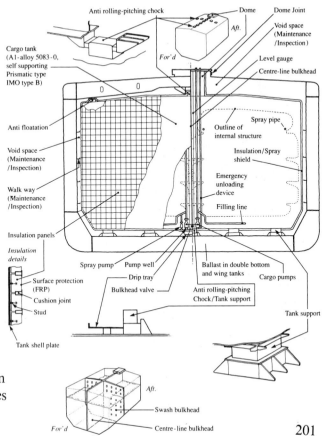

Fig. 16h. Cross section showing principal features of the IHI-SPB containment system.

201

Having obtained all the necessary Classification approvals by 1985, IHI were able to bid for new ships with complete confidence and happy in the knowledge that they were at no disadvantage to yards offering the Moss spherical design.

Finally, in 1989, after many years of dogged persistence, tests and computer runs, discussions with shipowners and Classification Societies, presentation of papers to the Gastech and other conferences, Dr. Fujitani, General Manager Gas Carrier Project, and his team were rewarded with an order for two 87,500m³ ships of the IHI-SPB design − the *Arctic Sun* and *Polar Eagle* − which will replace *Polar Alaska* and *Arctic Tokyo* when they complete their contract in mid/end 1993.

Fig. 16h shows details of the IHI containment system arrangement. Plate 2 shows details of the cargo tank internals, Plate 3 shows the first tank being lifted into position − the bottom supports can be clearly seen.

The IHI-SPB design has one attractive feature not possessed by the other three, namely the ability to subdivide the cargo tanks: this will increase in importance as a safety factor as individual tanks increase in size and, of course, it completely eliminates the sloshing problem.

So the trapezoidal independent LNG tank − upgraded by an enthusiastic and tireless design team, with the benefit of modern design tools and welding technology, and unfettered by 'restrictive' patents − lives to see another day and, judging by its predecessors, many years of trouble-free service.

Plate 2.   IHI Cargo Tank under
construction

Plate 3.   IHI Cargo tank en route
to building berth

## BOIL OFF REDUCTION

All containment systems have been able, by increasing insulation thickness and careful attention to details, to reduce the guaranteed daily boil off in the loaded condition to under 0.15%/day: this compares with 0.33% per day for *Methane Princess/Methane Progress* and 0.25% per day for the first 125,000m³ ships. Boil off on existing membrane ships can be reduced by creating a vacuum in the insulation space[9].

In some cases the boil off may be lower than that required to propel the ship and under these circumstances it may be necessary to 'augment' boil off − not too difficult with a suitable steam heater.

## CARGO HANDLING

During this decade of consolidation, the equipment installed for the handling of cargo, temperature and pressure measurement and monitoring has changed little in principle but has seen good progress in reliability − particularly in regard to instrumentation; here solid state has replaced mechanical switching with great benefits in reliability; ultrasonics are replacing mechanical floats in level gauging; tending to obviate the need for the ship's staff to 'pace the deck' during cargo transfer operations and ensure that all is going according to plan. This does not mean that convenient access to deck-mounted equipment is no longer important, but perhaps it is now a little less so.

The cargo pumps had always exceeded expectations in respect of both performance and reliability, and with ever increasing tank capacity and therefore pump size, this record has been maintained. A typical modern installation comprises two main pumps per tank, each of say 1,400m³/hr. capacity, tailor made for the capacity and head required for the project terminal and a 12 hour discharge, plus one emergency retractable pump of say 45m³ capacity in each tank. A good example of the submerged pumps fitted to modern LNG ships is illustrated in Fig. 16i manufactured by the Cryodynamics Division of Ebara International Corporation; it has a capacity of 1,650m³/hr (600 HP) compared with the 50 HP pumps fitted to *Methane Princess/Progress* in 1964. (Fig. 8c). An interesting, and beneficial, innovation is the 'thrust equalizing' feature which eliminates undesirable axial thrusts on the bearings. It will also be noted that the pumpage passes through an annular passage around (and within) the body of the pump, thus permitting an in-line mounting of the discharge pipe.

Submerged motor pump power cables on current production LNG carriers are quite different from the rigid mineral insulated cable system used on the earlier ships. Cables are now manufactured with stranded copper conductor, teflon insulation and covered with a braid of stainless steel and remain flexible at LNG temperatures. This type of cable also makes the procedure of lifting and lowering the emergency pumps very much easier.

Natural Gas by Sea

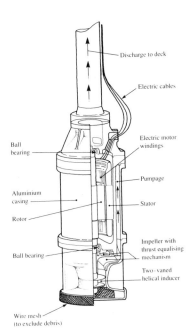

Discharge to deck

Electric cables

Electric motor
windings

Ball
bearing

Pumpage

Aluminium
casing

Stator

Rotor

Impeller with
thrust equalising
mechanism

Ball bearing

Two-vaned
helical inducer

Wire mesh
(to exclude debris)

Fig. 16i. Cutaway drawing
of typical Ebara-
Cryodynamics LNG
cargo pump.

**SAFETY**

The question of safety has always been in the forefront of LNG containment
designers' minds and the inherent safety margins available in new designs as
they come forward are thus of vital concern. Competition has, over the years,
undoubtedly exerted pressures towards reducing 'unnecessary' margins. The
Concise Oxford Dictionary defines margin as being 'the extra amount over
and above the necessary or minimum' ... but how precisely can we define
the necessary or minimum?

Safety standards set by the regulatory authorities are increasing in severity
year by year: designs which were developed in all good faith and considered
acceptable internationally twenty-years ago are no longer considered so. No
doubt this trend will continue – even though designers are daily increasing
their capability of greater precision – and this is in no small part due to the
greater awareness and appreciation of environmental risks.

Thus the early margins which catered for imprecision in design are being
replaced by stricter controls on designs of ever greater precision. So where
does that leave us? A gradual progression towards greater reliability and safety
– which is as it should be – the present excellent LNG safety record *must*
be maintained if the industry is to grow and prosper.

Whilst design and construction standards are catered for by IMO, who also
impose standards for staff and crew training, the Classification Societies and
USCG, the industry itself foresaw quite early the need for a Forum in which
ideas and information essential to the smooth operation of both ships and
terminals could be exchanged.

204

Thus the London based Society of International Gas Tanker & Terminal Operators (SIGTTO) was formed, largely at the initiative of the El Paso Group, with ten founder members, in 1979. The objectives, as stated in the Constitution, are as follows:

> "To protect and promote the mutual interests of its membership in matters relating to the safe and reliable operation of gas tankers and gas terminals, to advise and provide information to its members, to represent its members before, and consult with, the International Maritime Organization, other governmental bodies and the International Chamber of Shipping and Oil Companies International Marine Forum on matters relating to the 'terminalling' and shipment of liquefied gas."

By 1992 the Society had 84 members responsible for 75% of the world's LNG fleet, 56% of the LPG fleet, 90% of LNG terminals and 43% of LPG terminals. Just looking at these statistics is evidence enough of the deep concern of the LNG industry that the highest level of safety and reliability should be maintained in this rapidly developing and expanding field.

A list of SIGTTO's publications relating to LNG will be found in Appendix 5.

The subject of safety cannot be left without briefly discussing two matters: (1) the value of a double bottom in surviving grounding damage, (2) the risk of collision. A dramatic demonstration of grounding occurred in June 1979 when *El Paso Paul Kayser,* a 125,000m³ ship of Gaz Transport membrane design, ran aground at full speed (see p.206) in the Straits of Gibraltar with a cargo of 99,500m³ of liquefied gas.

The vessel remain fast on the rocks for four days sustaining widespread bottom damage extending over the entire 'cargo area'. She was refloated by the application of compressed air into her ballast tanks and towed to Algeciras Bay for the transhipment of her cargo to a sister vessel, *El Paso Sonatrach.* The cargo was transferred, using the ship's own pumps, through a single 10″ stainless steel bellows type hose, and completed within 48 hours[5].

Despite substantial local deformations of both inner hull and membranes, no leakage of cargo occurred, nor was there any sign of leakage of water into the containment spaces.

A second example was the case of *LNG Taurus* which grounded off Japan in December 1980, fully loaded and in bad weather − and on rocks. She was refloated without damage to the spherical containment system.

Had either ship been an oil tanker with a single shell there would have been a major pollution problem. Nevertheless one must admit that the owners of *El Paso Paul Kayser* were fortunate indeed that the seas remained calm between 30th June and 9th July 1979 − had the grounding occurred in bad weather the story might not have had such a happy ending. Refs[4/7].

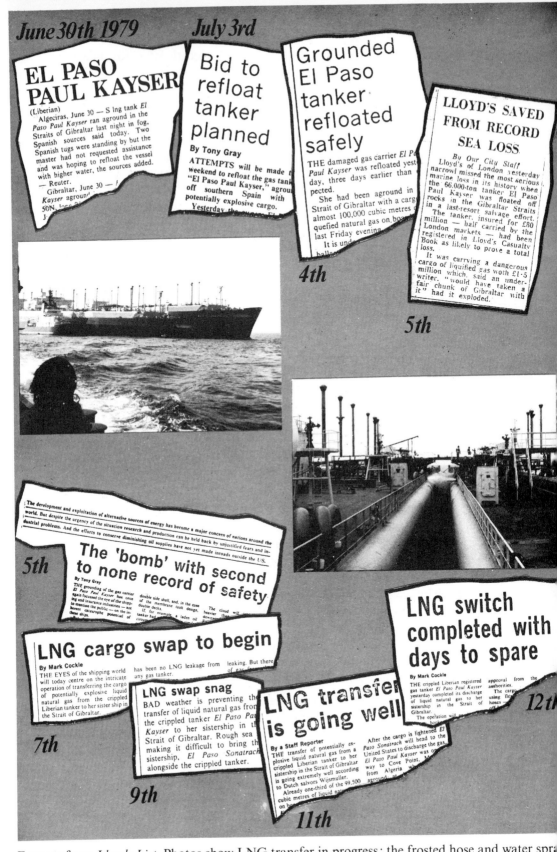

## EL PASO PAUL KAYSER

(Liberian)
Algeciras, June 30 — S lng tank *El Paso Paul Kayser* ran aground in the Straits of Gibraltar last night in fog, Spanish sources said today. Two Spanish tugs were standing by but the master had not requested assistance and was hoping to refloat the vessel with higher water, the sources added.
— Reuter.
Gibraltar, June 30 —
Kayser aground
50N, long

## Bid to refloat tanker planned

By Tony Gray

ATTEMPTS will be made t
weekend to refloat the gas tank
"El Paso Paul Kayser," agrou
off southern Spain with
potentially explosive cargo.
Yesterday th

## Grounded El Paso tanker refloated safely

THE damaged gas carrier *El Pa
Paul Kayser* was refloated yest
day, three days earlier than
pected.
She had been aground in
Strait of Gibraltar with a carg
almost 100,000 cubic metres
quefied natural gas on boa
last Friday evening
It is und
ball

*4th*

## LLOYD'S SAVED FROM RECORD SEA LOSS

By Our City Staff
Lloyd's of London yesterday
narrowl missed the most serious
marine loss in its history when
the 66,000-ton tanker El Paso
Paul Kayser was floated off
rocks in the Gibraltar Straits
in a last-resort salvage effort.
The tanker, insured for £80
million — half carried by the
London markets — had been
registered in Lloyd's Casualty
Book as likely to prove a total
loss.
It was carrying a dangerous
cargo of liquified gas woth £1·5
million which, said an under-
writer, "would have taken a
fair chunk of Gibraltar with
it" had it exploded.

*5th*

The development and exploitation of alternative sources of energy has become a major concern of nations around the
world. But despite the urgency of the situation research and production can be held back by unjustified fears and industrial problems. And the efforts to conserve diminishing oil supplies have not yet made inroads outside the US.

*5th*

## The 'bomb' with second to none record of safety

By Tony Gray
THE grounding of the gas carrier
El Paso Paul Kayser has once
again focussed the eye of the ship-
ing and insurance industries — not
to mention the public — on the inherent catastrophe potential of
these ships.

double side shell, and, in the case
of the membrane tank design,
double decks.
If, for example, a laden oil
tanker had
circum

The cloud will
heavier than
down

## LNG cargo swap to begin

By Mark Cockle
THE EYES of the shipping world
will today centre on the intricate
operation of transferring the cargo
of potentially explosive liquid
natural gas from the crippled
Liberian tanker to her sister ship in
the Strait of Gibraltar.

has been no LNG leakage from
any gas tanker.

leaking. But there
of gas

*7th*

## LNG swap snag

BAD weather is preventing the
transfer of liquid natural gas fro
the crippled tanker *El Paso Pa
Kayser* to her sistership in th
Strait of Gibraltar. Rough sea
making it difficult to bring th
sistership, *El Paso Sonatrach*
alongside the crippled tanker.

*9th*

## LNG transfer is going well

By a Staff Reporter
THE transfer of potentially explosive liquid natural gas from a
crippled Liberian tanker to her
sistership in the Strait of Gibraltar
is going extremely well according
to Dutch salvors Wijsmuller.
Already one-third of the 99,500
cubic metres of liquid na
on b

After the cargo is lightened El
Paso Sonatrach will head to the
United States to discharge the gas
El Paso Paul Kayser was on
way to Cove Point.
from Algeria w
aground

*11th*

## LNG switch completed with days to spare

By Mark Cockle
THE crippled Liberian registered
gas tanker *El Paso Paul Kayser*
yesterday completed its discharge
of liquid natural gas to her
sistership in the Strait of
Gibraltar.
The operation will
boo

approval from the
authorities.
The carg
using fle
hoses

*12th*

Extracts from *Lloyds List*. Photos show LNG transfer in progress: the frosted hose and water spra
'curtain', to protect the hull in the event of leakage, can just be seen.

That ship to ship transfer of LNG could be safety carried out even under adverse weather conditions was demonstrated both during this incident[5] and subsequently, when *LNG Libra* suffered a fractured tailshaft off Japan in October 1980[6].

Many studies have been carried out on the likelihood of, and risks involved in, collisions[10]. It is a matter of particular concern to port authorities and of course also to shipowners/operators, the insurance market and many others. Each of the containment systems can claim certain advantages in collision resistance. Does increase in ship size increase the risk? Much more is now known about what might happen in the even of LNG spills[11/12/13], both on land and water, but it is, nevertheless, a sobering thought that the entire contents of *Methane Princess* – and more – are now carried in one, single, undivided cargo tank.

How much further should we go in *this* direction in the search for reduced capital and operating costs?

## THE CONFERENCES

No book on LNG ship technology would be complete without some reference, however brief, to the conferences which provided and still continue to provide, an integral and essential forum for the regular exchange of information and the advancement of knowledge.

These conferences are:

(1) The Gas Industry sponsored International Conferences, now more familiarly known as LNG 1,2,3, etc, originally held every two years, now every three. They cover all aspects of the gas transportation field and include sections on the market, liquefaction, transportation, reception and distribution.

(2) The 'privately' sponsored, also international, LNG/LPG conferences, originally promoted by the marine journal *Shipbuilding and Shipping Record* and subsequently by Gastech Limited under the energetic leadership of Brian Singleton – one time editor of the aforesaid journal. These, now more familiarly known as the Gastech conferences, concentrate more specifically on the ship technology, although they by no means neglect the market place, the terminals, and being more commercially oriented and at the same time less formal, provide an immensely stimulating platform for the exhange of up to date information on technological progress and meeting the individuals concerned.

Many valuable contributions were made to the LNG conferences, with contributions readily available from the floor, but it was at the Gastech conferences, and their two predecessors in 1972 and 1973 that the sparks really flew! Furthermore, up to and including Gastech 86, the discussion periods are recorded in the Gastech proceedings and make fascinating reading in retrospect. These conferences covered the period of perhaps the most intense period of competition between the containment system designers in their struggle for 'supremacy' and none demurred from asking the most provocative questions of the other.

The reader will find many references at the end of each chapter to papers presented at both these Conferences; the proceedings form one of the few readily available sources of information on the art; their growth from 300 delegates at the 1972 LNG/LPG conference to 1,200 at Gastech 78 and 2,000 at Gastech 90 relates closely to the expansion of the industry.

It is hoped that the author will be forgiven for including here (Appendix 4a) a reprint of his introduction to the session on containment systems at the New York conference of 1976. It was an attempt to review the state of the art at that time in a not too boring fashion. Later, in 1984, another 'outburst' when the market was very sluggish (Appendix 4b).

[1] Repairs to the LNG Carrier "Tellier": Longevity of the Technigaz Technology by J. Claude & M. Etienne, Gastech 90, Amsterdam.

[2] Prediction of Sloshing Loads in LNG Ships by Dr. J.C. Peck and P. Jean Gastech 81, Hamburg.

[3] Technical and Economic Aspects of Gaz Transport LNG Carriers by R. Lootvoet, Gastech 85, Nice.

[4] Gaz Transport Technique, Experience and Performance by R. Lootvoet, Seoul Symposium, LNG & LPG-ROK, Seoul 1982.

[5] Ship to Ship transfer of LNG, the *El Paso Paul Kayser/El Paso Sonatrach* gas transfer operation, by F.E. Shumaker, Gastech 79, Houston.

[6] LNG transfer ship to ship following *LNG Libra* tailshaft failure. Masaitis & Tornay. Gastech 81, Hamburg.

[7] The Gaz Transport Membrane, a proven system combining Lower Operating costs, with a high degree of safety, by P. Jean & R. Lootvoet. Gastech 82 Paris.

[8] "LNG Log" published annually by SIGTTO.

[9] "Application of vacuums to the insulation spaces in order to reduce boil-off on membrane-type LNG Carriers", by Verschum, Wayne, le Nobel and Shibamura, Inst. of Mar. Engineers, London, 1985.

[10] "Studies of the resistance of LNG Carriers to collisions", by Capt. H.P.G. Neuner, M. Böckenhauer. LNG 6, Kyoto 1980.

[11] "Maplin Sands Spill Tests 1980" – Shell Publication.

[12] "US Coast Guard Liquefied Natural Gas Research at China Lake" A.L. Schneider et al. Gastech 79, Houston.

[13] "Shipboard jettison tests of LNG on to the Sea", by A. Kneebone, L.R. Prew. Gastech 74, Amsterdam.

# 17

# Projects Past, Present and Future

Every LNG project is a complex chain[1] extending from the well head, to the reception terminal, thence to the gas consumer. Each link of the chain, of which the shipping is just one, is part of an integrated contract which may cover a period of twenty-five years or more.

Investments are large and politics may also play a significant role[2].

In the early years some of the contractual snags were not fully appreciated and negotiations fell apart at a late stage; ships, ordered in anticipation, were then laid up — sometimes for many years.

The industry has experienced many such headaches in the past and has learned to live with them, and learn from them.

By the end of 1979 there were eleven projects in operation, involving 41 ships with capacities ranging from 25,000m³ to 125,000m³ and an investment in ships alone of over U.S.$30 billion. By the end of 1991 — just twelve years later — there were fifteen basic trade routes between the principle exporting and importing countries (Fig. 17a).

Not all these projects — or ships — were without their completion or start-up problems but they are all now in regular operation and provide dramatic evidence of the successful development of the technology involved.

As can be seen from Fig. 17b forecasters have been consistently over-optimistic, for example in February 1971, it was confidently stated "...that the demand for natural gas in the U.S. East Coast, in the face of declining ability of domestic resources to satisfy need, is such that it is safe to say that

209

| **Major LNG Trades – 1992** | | | |
|---|---|---|---|
| | | **Start-up Date** | **Ships(m$^3$)** |
| Abu Dhabi | — Japan | (1977) | 4 x 126,200; 1 x 87,600 |
| Algeria | — Belgium | (1987) | 2 x 125,000 |
| | France | (1965) | 2 x 125,000; 1 x 50,000; 2 x 40,000 |
| | Spain | (1969) | Var. $\equiv$ 3 x 40,000 |
| | U.K. | (1964) | 1 x 27,500 (part time) |
| | U.S.A. | (1978) | Var. $\equiv$ 2 x 125,000 |
| Australia | — Japan | (1989) | 8 x 125,000 |
| Brunei | — Japan | (1972) | 5 x 75,000; 2 x 77,500 |
| Indonesia | — Japan | (1977) | 8 x 126,300; 9 x 125,000; 1 x 86,000 |
| | Korea | (1987) | 2 x 125,000 |
| | Taiwan | (1990) | 1 x 136,400 |
| Libya | — Italy | (1969) | 2 x 40,000 |
| | Spain | (1971) | Var. $\equiv$ 2 x 40,000 |
| Malaysia | — Japan | (1983) | 6 x 130,000 |
| U.S.A. | — Japan | (1969) | 2 x 71,500; 2 x 87,500 from 1993 |

Fig. 17a

at least five major projects already much discussed will come to fruition in this decade...'' (Ref.[3]).

In fact only two came to fruition due largely to the political procrastination in the U.S.A. and one soon failed; but a number of ships were ordered in expectation and much embarrassment was subsequently caused by their enforced, and extended, unemployment.

An early 'role' model which provided a demonstration on how it should be done was the Shell Brunei-Japan project[4] initiated by Conch in 1964 but taken over in its entirety by Shell in 1967 after the other two shareholders had withdrawn their support.

It was an impressively smooth operation, both in the final stages of its negotiation – a computerized project simulation programme[5] based in London was used to provide overnight guidance to the negotiating team in Tokyo — and in the construction of the ships and liquefaction plant.

The whole project remained completely on schedule throughout its planning, construction and initial operating phases, every one of the seven ships being

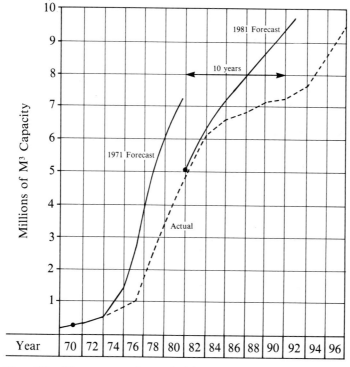

Fig. 17b. Forecasts of total ship capacity

delivered at the appointed time; they are still operating 'round the clock' 20 years later.

This was the first LNG project to be mounted on the grand scale and its undoubted success was due in no small part to Conch's solid groundwork in the LNG field since:

— the senior project director and negotiator for Shell was previously Conch's managing director;

— the plant construction manager at Brunei had been plant manager at Arzew;

— five of the seven ships were built to the Conch Ocean (now TGZ) membrane design;

— the fleet managers based in Tokyo had spent many years looking after *Methane Princess/Methane Progress* operation;

— several of the ship masters, chief engineers and cargo officers had previously sailed on *Methane Princess/Methane Progress*.

Plate 1. *Geomitra*–one of seven 73,500 m³ ships on Brunei-Japan run (Gaz Transport Membrane). Note arrangement for stern loading.

The only major project to have failed at a very advanced stage, indeed soon after its commencement in 1978, was the ill-fated El Paso Algeria-U.S.A. project[6]. This project was "by far the largest liquefied gas (LNG) transportation project in the world and the first to transport large quantities of Natural Gas into the United States by ship. The operating lifetime of the project is 25 years, a period extending into the 21st century. The project will provide energy sufficient to meet the average daily requirements of a metropolis two and one third times larger than New York City". Its demise was primarily due to disagreements between the two contracting parties on pricing, but the failure on trials of the first of the three Conch designed ships (May 1979) followed within a matter of days by the grounding of *El Paso Paul Kayser* at full speed off Gibraltar (June 1979) was hardly an auspicious start to what was, indeed, an LNG project on the grand scale; in fact, it collapsed in 1980. The three important terminals have been mothballed and six of the ships, after a prolonged period of lay-up, have been renamed and will be utilized elswhere.

The Abu Dhabi-Japan project (1977) was bedevilled in its early years by debris in the cargo transfer lines during the first ship loading, followed by failure of one of the shore storage tanks, but has subsequently run smoothly – plans are now in hand to expand and extend what has turned out to be a very successful contract.

212

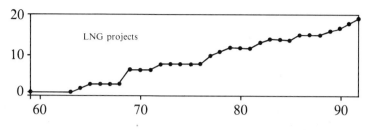

Fig. 17c. LNG projects

With these exceptions and a few of the normal teething problems experienced by any major undertaking, all current LNG projects are operating smoothly and efficiently and, it is understood, earning good money for the participants!

In addition to the fifteen projects listed above there are, inevitably, more waiting in the wings. As in the past, some will mature, some will not –

— like the El Paso project which so nearly succeeded;[6]
— like the efforts to develop marginal offshore gas fields worldwide, upon which Ocean Phoenix spent so much time and money;[7]
— like the grandiose Arctic Pilot project[8] which seemed all set to succeed in the late 1970s with two 140,000m³, 180,000 SHP Class 7 icebreaking LNG ships operating from Melville Island in the Arctic Circle to Quebec and the U.S.A.

General Dynamics Shipyard developed a submarine LNG tanker design for this project[9] which created tremendous excitement and enthusiasm amongst designers, consultants and shipyards alike – all of whom were convinced it would go ahead.

Some time in the future the market conditions will be right and this most imaginative project, upon which the Canadian Dome Petroleum Company expended so much time, effort and money, will assuredly be picked up, dusted down and implemented; most of those involved in its initial development will be sitting wistfully in their retirement homes, feet slippered, whisky in hand, relating to their grandchildren the great opportunity they were denied – but which they, their successors, will have the privilege of bringing to fuition. The author will be one of the former!

In contrast to those that failed to mature, the Nigerian export scheme, initiated by Conch in 1965, has at last achieved reality – a civil war and over 25 years later!

The Qatar-Japan project is expected to start up in 1997 and will require seven 137,500m³ ships – or their equivalent – to service its 25 year contract. To give the reader some idea of the huge investment involved in a major LNG

Which design will be used?

Melville photos.

Plates 2 and 3. Arctic Pilot Project.

214

project today, the cost of this one project is reported to be U.S.$3 billion for the liquefaction, storage and other shore facilities, including the loading terminal, plus $2 billion for the ships ($285 x 10[6] per ship); this does not include gas gathering and supply to the plant which could add a further $1 billion. Totalling six thousand million U.S. dollars — a lot of money, and requiring close attention to the small print in the contract![10].

Exports from Venezuela — first discussed as a potential supplier to the U.K. in the late 50s, only now, in the early 90s, are beginning to look like a 'goer' — to the U.S.A? Or Europe? Only time will tell.

It is an ever-changing scene — and a fascinating one.

[1] 'A systems approach to the Marine transportation of LNG by J.J. McMullen. International LNG Conference. Chicago 1968.

[2] A review of current LNG ship technology — and an attempt to rationalise'. R.C. Ffooks. LNG 5. Düsseldorf 1977.

[3] 'The pattern of sea-borne LNG trading in the 70's' Shipbuilding & Shipping Record, 29th Jan/5th Feb. 1971.

[4] 'The Brunei liquid natural gas plant' A.J.W. Ploum. LNG 5. Düsseldorf 1977.

[5] 'The role of operational research and computing techniques in the LNG business'. A.M. McCarthy & H.W. Walker. LNG 4 Algiers 1974.

[6] 'The First United States LNG base load Trade from Algeria — the Cove Point operation' Capt. J.W. Kime, J.W. Boylston, J. van Dyke, Gastech 79. Houston 1979.

[7] 'The OLASCO Offshore liquefaction and shipping systems for marginal gas fields'. K.W. Edwards, E.K. Faridany, J.E. Sloggett. Gastech 81, Hamburg 1981.

[8] 'Design options for an Arctic Class LNG Carrier,' R.A. Dick, V. Laskow, J. Wainwright, Gastech 79, Houston 1979.

[9] 'A submarine LNG tanker concept for the Arctic' Takis Veliotis, Spencer Reitz, Gastech 81, Hamburg 1981.

[10] LNG World overview 1991. Gotaas Larsen.

# 18
# A Look Ahead

"It is always difficult to predict, especially about the future."
(Danish proverb)

"Prospects are good for the next decade to become the golden age for LNG."
− Keynote speech, LNG 10 (1992).

"LNG industries worldwide have never faced a more welcoming market"
− Shell, LNG 10

"Today it is quite difficult to conceive of life without LNG."
− Tokyo Gas Co., LNG 10

"Optimization of LNG vessel costs is never done at the expense of existing safety standards."
− Gaz de France, LNG 10

"The rest of industry should emulate the high standards set by LNG terminal and ship operators."
− Shell, LNG 10

It is, perhaps, a little self-indulgent to subject the reader to the author's opinions on what the future might hold − but no book seems to be complete without a chapter on the subject!    This will be no exception.

As to the market place, it would appear from the pronouncements of those experts in day to day combat with the commercial side of the business that the future has never looked rosier, that there will be a 'very rapid expansion' (we don't use the word 'explosion'!) in demand − in all areas of the developed world − in the 1990s and beyond. After all, natural gas has always been a very environmentally friendly fuel and there is lots of it available. Its only drawback is the cost of production and transportation, since it must compete with oil, but one recalls that much the same was heard in 1971 and 1980 (Fig. 17b). Nevertheless, technology will respond to the challenge, as it has in the past.

The present challenge is, in simple terms to *reduce* transportation costs *without any sacrifice* in the high standards of safety and reliability which the industry has established to date. Of course, this challenge has been ever present; the membrane design was developed in 1960 with the sole purpose of reducing the capital cost of the ship − by some 25%! The membrane succeeded, but the cost advantage didn't quite materialise. Similarly the Moss sphere was developed in 1970 as a means of virtually eliminating the expensive secondary barrier requirement − in this latter it did succeed, but the overall cost advantage was not great; the Conch PUF insulation system was introduced only to reduce the high secondary barrier cost − and failed.

For some years now the step by step approach has been adopted by all the three main containment systems and it is most likely that this approach will continue for the foreseeable future. In any case, the investment in LNG projects today is now so high, and the cost of the failure, or even partial failure, of any one of the links in the chain so great, that any major departure from well established principles and practice is almost impossible to conceive.

Developments which can be clearly foreseen are

(1) **Increase in individual cargo tank size** − i.e. by reducing the number of cargo tanks in the traditional 125,000m³ size tanker from five to four, each then having a capacity of 32,500m³ (well over the total capacity of the first commercial tankers!). This arrangement has already been adopted by the Moss spherical system whose four-tank ships are now in operation on the Australia-Japan run; and is incorporated in the first two LNG ships to have been built by a Korean shipyard − Hyundai, as illustrated in Plate I. Approvals have also been obtained for these larger tank sizes for both membrane designs.

For a sphere this means a 9% increase in diameter, a 5% overall reduction in surface area, for the four larger tanks compared to the five smaller tanks and a 30% increase in individual tank capacity, leading to lower boil off, simpler cargo handling system and an undisclosed net saving in capital and operating cost. Similar cost advantages will apply to the membranes and also to the more recently

217

Plate I. Artists impression of first LNG ship built in S. Korea

re-introduced trapezoidal tank system. In this latter, however, the individual compartment size would be 16,250m³.

Assuming no problems develop in the tank support structure this trend may well continue − unless and until some Port Authority becomes 'apprehensive' about the potential spill size in the event of a collision.

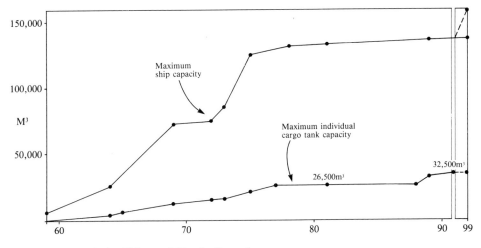

Fig. 18a. Trends in Ship and Tank Capacity

(2) **Larger Ships** — It is difficult to know why a significant increase in ship size has not yet taken place, indeed why it did not happen in the late 70s or early 80s; the technology has been in place since the mid 70s, when almost every LNG ship builder had designs available for vessels of up to 200,000m³ capacity.[5]

The economics do not appear to have been in dispute; for example, moving from 125,000m³ to 165,000m³ shows a 10% saving in transportation cost for a typical project[5/6], and this includes the costs of additional land storage at the loading and discharge terminals. Now that important terminals, certainly in Japan, have very large storage capacities, this element can perhaps be partly discounted.

Potential problems are (1) draught, which may present difficulties in some locations above 165,000m³, but this can surely be overcome by an imaginative naval architect, and (2) jetties, presently designed for 125,000/135,000m³ size, would in many cases require modifying for the larger vessels — not a great cost by comparison with the operational savings over a twenty-five/thirty year contract or longer.

Bearing in mind the future potential congestion facing LNG ships entering Japanese waters, and particularly Tokyo Bay, it must make sense to employ fewer larger vessels. But who will be the one to break away from the present 125,000/135,000m³ ship? Not, after all, a very fundamental step.

(3) **Diesel propulsion and reliquefaction of boil off**[1/2/3/7] — To date virtually every commercial LNG ship has been propelled by dual fuel steam turbine machinery with boilers fired primarily on boil-off gas; burning oil fuel exclusively when manoeuvring into and out of port. Natural gas is a clean and readily available fuel and the dual fuel burning technology is now well known and reliable.

Much study has been given by the larger diesel engine manufacturers to the development of dual fuel marine diesels. As so often there are pros and cons. Gas must be supplied to them at very high pressures (2,500psi compared to 20psi for boilers) hence the requirement for high duty compressors (extra maintenance). Their low fuel comsumption would, in most cases, not utilize the total boil off — hence whole or partial reliquefaction of the boil off would be needed (more equipment and maintenance). In their favour is the greater availability of engineering staff familiar with diesel machinery — almost every other type of ship at sea is propelled by diesel machinery — and fuel economy.

219

It seems very probable, particularly for those new projects where round trip distances are high — Qatar-Japan: 12,500 n.m., Nigeria-U.S.A.: 10,000n.m., compared to Indonesia or Australia-Japan: 6,000/7,000n.m. — that the saving in cargo loss due to boil-off (a premium fuel) combined with greater staff availability may tip the scales in favour of diesel propulsion. It will not be a major technological step.

(4) **Increased ship life span** — It will be recalled that early projects were based first on a fifteen year and later twenty, then twenty-five years ship life span. As these contracts drew to a close it became clear that, provided they had been well maintained, the physical life of the ships had been under-estimated.

*Methane Princess,* built in 1964 for a fifteen year contract, was still in active service twenty-five years later. *Polar Alaska* and *Arctic Tokyo,* built for a fifteen year contract commencing in 1969, extended by five years in 1982, will be replaced by two larger vessels in 1993 — twenty-five years later — not to be scrapped but bought "in excellent condition" by British Gas for further service — presumably between Algeria and the U.K.?

Similarly the operating lives of the Moss spherical ships on the Abu Dhabi-Japan run, whose twenty year contract is due to expire in 1996, are being reassessed; likewise the seven membrane ships on the Brunei-Japan run.

Each ship and design will be assessed on its past performance and present condition but ship lives of thirty and forty years are now being very seriously — and realistically it would seem — considered[4]. LNG is non-corrosive, so that operating in an inert gas or dry air atmosphere, with stainless steel piping and deck equipment, the 'cryogenic parts' appear to have an almost infinite life. A potential weak link is the ship's inner hull — subjected to corrosive action from the ballast against its outer surface — but epoxy coatings, cathodic protection and vigilant supervision seem to have this factor under control. Modern coatings, continuous machinery surveys and maintenance will deal with the rest of the ship! Technological obsolescence is the only remaining problem.

(5) **New containment systems** — It is difficult to see how any radically new containment system can now succeed. Quite apart from the very considerable time and cost to develop, it must as always hold out a potential capital cost saving of well over 20% to have any chance at all. This possibility now seems so remote as to be virtually non-

existent.

The 'ultimate' design must be the one in which some magical indestructable insulation system is sprayed on to the internal surfaces of the steel holds of the ship and contains the liquid cargo, with no sign of deterioration, for thirty or forty years. At one time the advent of polyurethane foam seemed to put this goal almost within our grasp, but no sooner was it put to even a modest test than it failed completely! Now the ultimate seems as far away as ever − perhaps we can include it in the tenth edition of this book by which time the author will be gazing down from above − still watching progress with keen interest.

On that salutary note another chapter in the history of the development of LNG technology − and this book − will quietly close.

[1] 'Boil off reliquefaction facilities for Diesel-driven LNG Carriers' E. Berger. Gastech 82, Paris.

[2] 'A New Generation of LNG Carriers for Economy and Operational Flexibility' R. Ogiwara et at Gastech 85, Nice.

[3] 'Development of the Sulzer Dual Fuel Diesel engine' B. Engersen et al. Gastech 86, Hamburg.

[4] LNG 10, Session IV, 5 papers on 'Longevity of LNG Carriers'.

[5] 'The Conch 165,000m$^3$ LNG tanker' R.C. Ffooks. Gastech 74, Amsterdam.

[6] 'Chantiers d'Atlantique and LNG transportation' discussion paper. Gastech 78, Monaco.

[7] 'Propulsion of LNG Carriers by Steam Turbines — a necessity or a Tradition' by Roger Courtay, LNG 10, Kuala Lumpur, 1992.

# Appendices

# 1
# Appendix
# What is Natural Gas?

Natural Gas is a mixture of hydrocarbons in which methane predominates; the mixture varies considerably according to where and how the gas is found – and particularly according to whether it is 'associated' with a crude oil reservoir – or 'unassociated'.

Associated gas is 'richer' because it contains a higher percentage of ethane ($C_2H_6$), propane ($C_3H_8$), butane ($C_4H_{10}$), all of which have a higher calorific value than methane ($CH_4$).

Natural gas may also contain a small percentage of nitrogen.

*Methane* itself cannot be liquefied by pressure alone unlike ethane, propane and butane; it must therefore be cooled in order to convert it into a liquid for marine transportation. It is normally cooled to its atmospheric boiling point of $-161.5°C$, or close to this temperature, for economic shipment, and in this condition occupies 1/600th of its volume as a gas.

The boiling point of a natural gas mixture will often differ somewhat from that of pure methane, being influenced by the boiling points of the other gases in the mixture.

Liquefied natural gas (LNG) is colourless, odourless and non-toxic; it has a relatively low flame speed, high percentage flammability level in air, high auto-ignition point and is non-corrosive; its weight is less than half that of water. Its main constituent, methane, is lighter than air at all temperatures above $-110°C$ (butane, propane and ethane are heavier-than-air gases at all temperatures).

Were it not for its very low temperature, LNG would be a relatively benign product compared with, say, gasoline.

223

# 2

# Appendix
# Some Approximate
# Physical Properties
# of Constituents
# of Natural Gas

# Some Approximate Physical Properties
# of Constituents of Natural Gas

|  | Methane | Ethylene | Ethane | Propane |
|---|---|---|---|---|
| Molecular formula  ..  ..  .. | $CH_4$ | $C_2H_4$ | $C_2H_6$ | $C_3H_8$ |
| Molecular weight  ..  ..  .. | 16.04 | 28.05 | 30.07 | 44.09 |
| Specific heat, btu/lb/°F ..  .. | at -175° = 0.4502<br>at -112° = 0.5308<br>at 59° = 0.5284<br>50°-392° = 0.5931 | at -131.8 = 0.3086<br>at 59° = 0.3592<br>59°-212° = 0.399<br>77°-392° = 0.430 | at -115.6° = 0.3475<br>at 59° = 0.3861 |  |
| Boiling point, °F(°C).. .. .. | -258.7 (-161.5) | -155.2 (-104) | -128.2 (-89) | -43.87 (-42) |
| Melting point, °F.. .. .. .. | -299.2 (-184) | -273.1 (-169.5) | -297.8 (-183.2) | -309.82 (-190) |
| Liquid density at bp .. .. .. | 0.415 | 0.566 | 0.561 | 0.585 |
| Vapour density, sp gr, air = 1 | 0.5544 | 0.9749 | 1.0493 | 1.554 |
| Heat of vaporization, btu/lb..<br>(kcal/kg) | 220.5<br>(122.5) |  | at 32° = 135.0<br>at -4° = 156.6<br>at -40° = 175.5<br>at -130° = 228.6 | at 68° = 150.12<br>at 30° = 161.28<br>at -22° = 176.40 |
| Flammable limit in air, per cent:<br>Lower  .. .. .. .. .. .. ..<br>Upper  .. .. .. .. .. .. .. | <br>5.00<br>15.00 | <br>2.75<br>28.6 | <br>3.00<br>12.50 | <br>2.12<br>9.35 |
| Relative volumes as liquids at<br>boiling point and gas at 70°F | 630 | 485 | 488 | 316 |
| Critical temperature °F (°C) | -116.5 (-83°C) | 50 (10) | 90 (32.2) | 206 (96.67) |
| Latent heat, btu/lb at bp  ..<br>(Kcal/Kg)  .. .. | 219.7<br>(123) | 207.56<br>(116.2) | 210.7<br>(118) | 183.5<br>102.7) |

# APPENDIX 2

## Some Approximate Physical Properties of Constituents of Natural Gas

|  | Butane | Nitrogen | Air |
|---|---|---|---|
| Molecular formula .. .. .. .. .. .. | $C_4H_{10}$ | $N_2$ |  |
| Molecular weight .. .. .. .. .. .. .. | 58.12 | 28.02 | at 212° = 0.2404 |
| Specific heat, btu/lb/°F .. .. .. .. .. |  | at -293.8 = 0.256<br>at 59° = 0.2477 | at 752° = 0.2430 |
| Boiling point, °F(°C) .. .. .. .. .. .. | + 31.1 (-0.5) | -320.8 (-196) | -127.3 (-88.5) |
| Melting point, °F(C°) .. .. .. .. .. .. | -211 (-135) | -345.9 (-210) | -227.6 (-144.2) |
| Liquid density at bp .. .. .. .. .. .. | 0.600 | 0.808 | 0.874 |
| Vapour density, sp gr, air = 1 .. .. .. .. | 2.0854 | 0.9672 | 1.00 |
| Heat of vaporization, btu/lb .. .. .. ..<br>(kcal/kg) | 164.7 | 85.68<br>(47.6) | 91.75 |
| Flammable limit in air, per cent:<br>Lower .. .. .. .. .. .. .. .. ..<br>Upper .. .. .. .. .. .. .. .. .. | <br>1.86<br>8.41 |  |  |
| Relative volumes as liquids at<br>boiling point and gas at 70°F .. .. .. | 108 | 690 | 760 |
| Critical temperature °F (°C) | 306 (152.2) | -232.8 (-147.1) | -221.3 (-140.7) |
| Latent heat, btu/lb at bp .. .. .. .. .. ..<br>(Kcal/Kg) .. .. .. .. .. .. | 165.9<br>(93) |  |  |

226

# 3
## Appendix
US Coast Guard Tentative Standards for Transportation of Liquefied Inflammable Gases at Atmospheric Pressure as revised by the API Special Committee on the Transportation of LNG by Water to **August 22, 1956** and edited by the Secretary of the Committee

## APPENDIX 3

**General Principle**
The proposed Methods of Transportation will not create safety hazards in excess of those normally encountered in the water movement of Grade 'A' inflammable liquids or liquefied petroleum gases under pressure.

**Section A – Design of Cargo Tanks**
The Committee uses the following definition for this discussion:
(a) Integral type tanks are part of a ship's structure.
(b) Independent type tanks are not a part of a ship's structure and are designed to carry the load imposed by the cargo.
(c) A liner is a non-load carrying membrane which separates the cargo from the insulation and is suitably retained in position at all times.
(d) A secondary barrier is an arrangement or structure designed to contain the cargo temporarily if a leakage develops in a primary container.

A-1   The design of cargo tank structure is subject to the approval of the American Bureau of Shipping or other recognized classification societies.

(a) Integral Tanks
Integral tanks shall be designed in accordance with classification society rules for vessels intended to carry oil in bulk.

(b) Independent Cargo Tanks
Independent cargo tanks are to be designed for a head at least equal to the height of the cargo hatch above the top of the tank plus 4 feet, or to the highest level that the liquid may attain or 2.0 psig, whichever is greatest, using the specific gravity of product(s) to be carried. The design criteria shall consider the forces and conditions which the tanks may be subjected to while in *tests* or in *service*.

A-2   Independent tanks shall be segregated from the main hull of the vessel.

A-3   (a) All cargo tanks vented at a gauge pressure of 4 psig or less shall be constructed and tested as required by standards established by the American Bureau of Shipping or other recognized classiffication societies.
All cargo tanks vented at gage pressures exceeding 4 psig but not exceeding 10 psig pressure will be given special consideration by the Commandant.

(b) Cargo tanks shall not be subjected to pressure to discharge them in excess of that for which the relieving device is set.

228

A-4    Butts and seams of externally insulated cargo tanks shall be full penetration welds. Such joints as may be required shall be examined by acceptable methods of non-destructive tests.

A-5    Structural members of the vessel not suitable for use at the temperature of the cargo shall be protected by adequate means from being cooled by the cargo to temperatures below that for which the material of such members is suitable. For such structural members adjacent to the cargo space, an acceptable system shall be provided for detecting abnormally low temperature.

A-6    A secondary barrier shall be provided where leakage from the primary container may cause lowering of the temperature of the ship's structure to an unsafe level.

**Section B–Installation of Cargo Tanks**

B-1    Independent tanks shall be supported on substantial foundations. Tank supports shall be arranged so as to avoid excessive concentrations of loads on the supporting portion of the tanks. Independent cargo tanks may be installed in insulated holds or compartments. Provisions are to be made for the thermal movements of tanks.

B-2    Installation of tanks and arrangement of insulation are to be such as to permit access to one side of a bulkhead, structure, or tank plating.

B-3    All cargo tanks shall be entered directly from the weather deck.

**Section C  –Insulation**

C-1    Insulation shall be suitably protected against penetration of moisture.

C-2    Insulation shall be resistant to the cargo. Insulation shall be properly supported and shall not deteriorate in normal service.

C-3    Where insulation is in contact with the cargo the method of installation and the material must be proved by acceptable tests to be suitable for the cargo at the service conditions.

**Section D–Mechanical Equipment, Piping, Valves, Fittings and Accessories**

D-1    All such items shall be of design and material suitable for the temperature, pressure, and service encountered and resistant to attack by the cargo carried.

D-2    Piping may be joined with butt welds, with flanged joints kept to a minimum. Socket and slip on connections may be used for sizes 2 inches and smaller. Threaded joints may be used for sizes one inch and smaller. When threaded joints are used, they shall be visible and accessible for inspection under all service conditions and limited to accessory lines properly valved off from the main lines, except that relief valves shall not be valved off from the main lines. Where joints are threaded and sealed by welding or brazing they need not be exposed.

D-3  Provisions shall be made to protect piping from excessive stresses due to temperature changes, and/or movement of tanks and equipment to which the piping is attached. Expansion joints shall be held to a minimum and where used shall be subject to approval.

D-4  Piping shall be adequately secured and lines that may be subject to cargo temperature shall be provided with temperature isolation from the ship's structure. Drip pans, covers, or other suitable protection for structural members shall be provided where cargo leakage and spillage may occur.

D-5  Piping, generally, shall enter the cargo tanks above the weather deck. Piping below the weather deck is subject to special study and approval. If piping is below the weather deck the piping must be in a compartment suitable as a temporary cargo container which shall be vented and entered from above the weather deck.

D-6  Independent filling and discharge connections will be required at the point where the piping enters the tank and within the tank unless provisions are made for preventing excessive localized cooling when the tank is being filled without the need of such a piping arrangement.

D-7  Suitable evidence shall be provided that the operating procedures for filling and emptying the cargo tanks will not result in excessive stresses in the tanks.

D-8  A liquid level gaging device shall be provided in each tank to determine the level of the liquid without opening the trunk. Independent high level alarms are required.

D-9  Manually operated shut-off valves shall be provided on all connections to each tank except connections to safety relief devices. Such valves are to be located as close to the tank as possible.

D-10  A relief valve shall be provided to prevent excessive pressure in any section of a line that may be blocked off. This relief valve may discharge into a vent header or into a line to a reliquefaction header. This reliquefaction header shall have a pressure relieving device.

D-11  Cargo pumps and compressors shall be provided with suitable shaft or rod seals.

D-12  (a) A venting system shall be provided consisting of a branch vent line from each tank connected to a vent header. Consideration shall be given to the vapors formed by heat from the walls of the tanks while they are being cooled.

(b) The capacity of branch vents or vent headers shall depend upon the number of cargo tanks connected to such branch or header capacity as provided for in the table 38.20-1(b), and upon the total relief valve discharge capacity. Table from 38.20-1(b) Capacity of Branch Vents or Vent Headers.

| No. of Cargo Tanks | Percentage of Total Valve Discharge |
|---|---|
| 1 or 2 | 100 |
| 3 | 90 |
| 4 | 80 |
| 5 | 70 |
| 6 or more | 60 |

D-13 Vacuum relief valves shall be provided on each tank to operate at a differential pressure within that for which the tank is designed. They shall have a capacity determined by the maximum discharge rate of the pumps.

Vacuum relief valves shall be designed, constructed and flow tested for capacity in conformance with Subpart 162.017 of Subchapter Q (Specifications). Vacuum relief valves shall be attached to the tank near the highest point of the vapor space. Shut-off valves shall not be installed between the tanks and vacuum relief valves.

Each vacuum relief valve shall be tested in the presence of an inspector before being placed in service. The tests shall satisfactorily indicate that the valves will start to discharge at a pressure not in excess of the maximum allowable pressure of the tank.

D-14 The maximum liquid loading rate shall not exceed that for which the venting system is designed.

D-15 The vent header riser shall be carried to a minimum height of one-third of the beam of the vessel above the weather deck. Means shall be provided to prevent rain or snow from entering the venting system. The vent discharge riser shall be so located as to provide protection against mechanical injury. No valve of any type shall be fitted in the vent pipe between the vacuum relief valve and the vent outlet. Suitable provision shall be made for draining condensate which may accumulate in the discharge pipe.

D-16 Special consideration shall be given to the venting system of tank barges operating on the inland waters where vent risers may be damaged by obstructions.

D-17 All tank inlet and outlet connections, except safety relief valve liquid level gaging devices, and pressure gages shall be labelled to designate whether they terminate in the vapor or in the liquid space. Labels of corrosion-resistant material may be attached to valves.

## Section E – Pressure Relief Valves and Relieving Devices

E-1   Cargo tanks shall be fitted with approved pressure relieving valves for operating conditions to vent vapors so that the pressure within the tanks will not exceed the pressure for which the tank is designed. This pressure relieving valve shall have the capacity to handle the vapors formed by the heat transfer into the cargo tanks within an ambient temperature of 115°F plus the vapor displaced by the maximum loading rate.

Pressure relief valves shall be designed, constructed and flow tested for capacity in conformance with Subpart 162.017 of Subchapter Q (Specifications). Pressure relief valves shall be attached to the tank near the highest point of the vapor space. Shut-off valves shall not be installed between the tanks and pressure relief valves.

Each pressure relief valve shall be tested in the presence of an inspector before being placed in service. The tests shall satisfactorily indicate that the valves will start to discharge at a pressure not in excess of the maximum allowable pressure in the tank.

E-2   The relieving valves shall have a total relieving capacity to handle the vapors formed by fire exposure to the walls of the cargo tank computed by the formula:

Q equals 21000FA 0.82

where   Q equals total BTU per hour absorbed by the tank

A equals 35 per cent of the vertical side wall area of the tank in square feet

F equals 0.5 where a metal screen wall is located between the radiant source of heat and the tank wall

In no case shall the pressure relief valve be less than 4 inches in diameter.

## Section F  – Tests and Inspections

F-1   All cargo tanks shall be tested after installation with a head of water as follows:

(a) Integral tanks shall be tested as required by American Bureau of Shipping or other recognized classification society rules.

(b) Independent tanks shall be tested to a height 4 feet above the hatch cover.

F-2   (a) Periodic Tests

Periodic tests shall be in accordance with classification society rules for tanks intended to carry oil in bulk.

     (b) Periodic Inspections

Each tank shall be subjected to an internal examination biennially at the inspection period.

# 4a
# Appendix

## LNG — the inside story

### I

There once was a Company 'C'
who wanted to ship LNG;
they put it in box-shaped aluminum tanks
with a balsa-wood second'ry B . . .
. . . then filed lots of patents and sat back with glee

Then along came our friend Monsieur 'B'
with a wonderful plan
for a waffled tin can
and a similar one for the second'ry 'B'
to the angry reaction of company 'C'

Monsieur 'GT' thought similarly,
though his tin cans were flat—
and no worse for that—
and so was his second'ry 'B'
—to the increasing fury of company 'C'
and wrath of our friend Monsieur 'B'.

Then out of the blue Messrs. 'M' and 'NV'
conceived the idea
of supporting a sphere
on a 'skirt', or a ring
and the whole bl---y thing
made so beautifully
that it needed no second'ry 'B'
—to the teeth gnashing, foot stamping, arm
waving fury of companies 'C', 'B' and 'GT'.

### II

So the shipowners bought lots of tin cans and
    spheres,
saved lots of money and sailed out to sea;
but strange to relate, though the tin cans still work
and the boxes have paid off their capital fee,
the spheres in their splendour—
—and low shipyard tender,
have ne'er shipped a cargo of liquid NG.

### III

And still they come into the fray,
a new design almost each day;
reinforced concrete, semi-membranes and lobes,
cylinders, wetwalls and polyfoam triax—
all of course blessed by the Coastguard and IACS.
But which is the cheapest? The best? Will last
    longest?
Which of the shipowners' nerves are the
    strongest?
How on earth will they get past the dread FPC?
(perhaps they should all join the Baltyk TC?)

### IV

My only advice faced with problems like these
is to gaze towards Heaven and pray—
GOD HELP US—PLEASE.

"presented as introduction to transportation and technology session at Gastech 76, New York"

# 4b
# Appendix
## from Gastech 84 proceedings

*The Boyhood of LNG — with apologies to Millais*

## The story of this picture goes like this:

They met him sitting on the quay
— just sitting, staring out to sea.
"What's wrong, old man?" "Tis sad" said he,
"There once was hope — expectancy,
many designs and a building spree,
a tremendous amount of activity,
but now ...
... come sit and listen carefully
and I'll tell you what happened to LNG",

"It all began in '59,
when the world was short of energy
and the one great hope for industry
was vast supplies of LNG."
"Hundreds of people — including me,
designed new ships, as you'll soon see.
Projects were mounted,
profits were counted,
ships were built,
and tears were spilt...

"And suddenly —
a fleet of vast capacity
to import gas to the USA
was conceived and built — or so they say,
for not a trace of it's left today."

"There were twenty-two different designs", said he
"All fully tested and ready for sea;
and the sad thing is, after all this time
we've ended up with only three."

"Which are those?" asked the boys expectantly,
"Be quiet" said the man, "And I will tell thee".

"Membranes" — he shuddered — "t'was plain to see
were cheaper than boxes of 5083"
(unless they're re-classed Independent Type B).
"Spheres" — at the mention of which he seemed
close to tears—
"They came along with 3-D, FE".

"Are they the best then?" asked the boys,
"Keep quiet" said the man.

"There are lots in service and doing well,
They make lots of money, but look like hell,
They blow in the wind and the Master can't see
but there'll be many more, believe you me."

"Are gas ships safe?" asked the boys
"if we decide to go to sea?"
"Yes, of course", said the man "as safe as can be".
"We're off" said the boys, "Good luck" said he,
So they left him staring out to sea.

# 5

# Appendix
# List of SIGTTO
# Publications relating
# to LNG

**Title**

1 Safe Havens for Disabled Gas Carriers. A Consultative Document in the Seeking and Granting of a Safe Haven. November 1982

2 Recommendations for the Installation of Cargo Strainers on LNG Carriers. January 1984

3 Safety Aspects of the Marine Transportation and Storage of Refrigerated Liquefied Fuel Gases — A Review of the State of the Art. January 1985*

4 Information Paper No. 1: IMO Status of Codes, Protocols and Conventions applicable to Liquefied Gas Carriers. January 1987*

6 Information Paper No. 3: The Controlled Dispersion of Liquid Spill and Vapour Emission Incidents by Water Spray. November 1987

7 Information Paper No. 4: The History of Incidents in the use of Hoses, ERS Dry-Break Couplings and Marine Loading Arms in the Ship-to-Shore and the Ship-to-Ship Transfer of Liquefied Gases. November 1987*

8 Information Paper No. 5: Ship/Shore Interface Communications. November 1988

9 Information Paper No. 6: Report of a Working Group on Liquefied Gas Sampling Procedures. November 1988

10 Information Paper No. 7: An Index to the IMO International Gas Carrier Code. October 1989

11 Information Paper No. 8: Human Error. May 1990

12 Information Paper No. 9: Human Error: Error Reduction Strategies. January 1991

13 Information Paper No. 10: Human Error in Communications: The Potential for Misunderstanding. October 1991

14 Information Paper No. 11: Total Quality Management — a Quiet Revolution. May 1992

15 An Information Note on Training Courses available for Bridge Team Training and Bridge Procedures, Liquefied Gas Cargo Handling (for Ship and Terminal Operators), Firefighting — Gas/General Fires. May 1992*

16 Guidelines for Hazard Analysis as an Aid to Management of Safe Operations. May 1992                                                                              W. & Co.

17 LNG Log — A Record of Voyages completed by the World Fleet of LNG Carriers (Produced annually)

19 Prediction of Wind Loads on Large Liquefied Gas Carriers. (OCIMF/SIGTTO). July 1985                                                                              W. & Co.

21 Textbook: Liquefied Gas Handling Principles on Ships and in Terminals. March 1986 (Available in English, Japanese and Spanish).                             W. & Co.

22 Cargo Firefighting on Liquefied Gas Carriers                16mm    film
   (Complete with Study Notes). May 1986                         Video

23 Study Notes of above title, available separately.                        W. & Co.

24 A Guide to Contingency Planning for the Gas Carrier Alongside and Within Port Limits. (ICS/OCIMF/SIGTTO) April 1987                                    W. & Co.

25 Recommendations and Guidelines for Linked Ship/Shore Emergency Shut-Down of Liquefied Gas Cargo Transfer. July 1987                                   W. & Co.

26 Guidelines for the Alleviation of Excessive Surge Pressures on ESD. July 1987                                                                              W. & Co.

27 A Guide to Contingency Planning for Marine Terminals Handling Liquefied Gases in Bulk. (ICS/OCIMF/SIGTTO) January 1989                          W. & Co.

28 A Contingency Planning and Crew Response Guide for Gas Carrier Damage at Sea and in Port Approaches. (ICS/OCIMF/SIGTTO 2nd Edition August 1989
                                                                                   W. & Co.

29 Inspection Guidelines for Ships Carrying Liquefied Gases in Bulk. (OCIMF/SIGTTO) September 1990                                                      W. & Co.

30 Ship Information Questionnaire for Liquefied Gas Carriers. (OCIMF/SIGTTO) September 1990.                                                                  W. & Co.

   * Publications restricted to members only.

W. & Co. Publications available through:-
Witherby & Co. Ltd.,
32-36 Aylesbury Street
London EC1R 0ET

Tel No: 071-251 5341
Fax No: 071-251 1296

# Name Index

# INDEX

# Subject Index

# INDEX